Physics-Rule Table Zyn 2

© 2023 Diogo de Souza
All Rights Reserved.
Contact Information:
diogodesouza7@gmail.com
diogodesouza7@hotmail.com

Table of Contents:	Page
Worksheets:	3

PreTest Review:	3
Work Power Energy	10
Momentum and Impulse	87
Thermal	141
Oscillations	163
Gravity and Electromagnetism	260
Nuclear Physics	459
Extra Problems	494
Sound Notes	542
Final Exam	543
Game	555

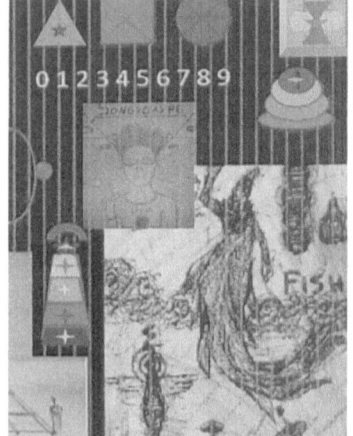

Name_____ Period_____

Pre Test Review

In this assessment you are not allowed to use a calculator and will be tested in your mathematical skills.

1. Using the Cartesian Coordinate System answer the following questions:

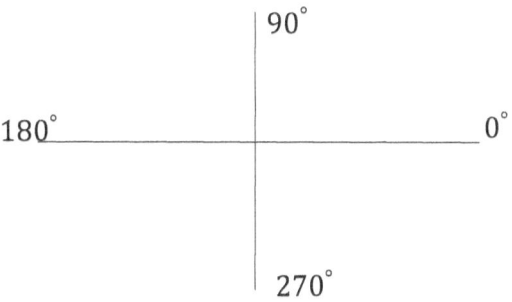

A. $\sin(0) =$
B. $\cos(0) =$
C. $\sin(90) =$
D. $\cos(90) =$
E. $\sin(180) =$
F. $\cos(180) =$
G. $\sin(270) =$
H. $\cos(270) =$
I. $\sin(360) =$
J. $\cos(360) =$
K. $\tan(0) =$
L. $\tan(180) =$

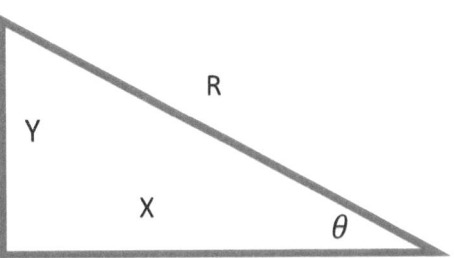

Trigonometry and the study of angles was essential to mapping the night sky and to understand the visible universe in the ancient world.

2/ Two common triangles.

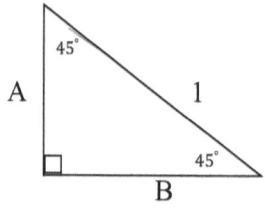
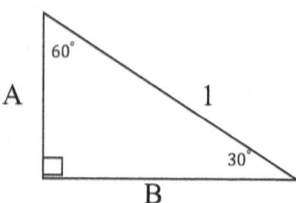

Find the value of:

A=
B=

A=
B =

2. Prove using the sine and cosine functions the Pythagorean Theorem:

Proof:

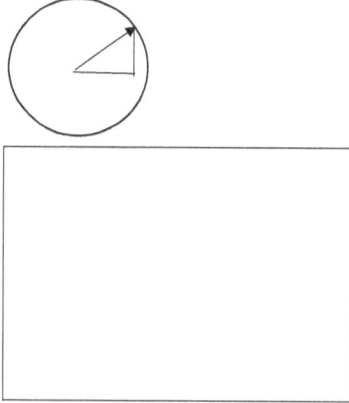

Pythagoras popularized his theorem in Greece after many trips to Babylon and Egypt.

3. Algebraic Manipulation:

Simplify the following:

A. $\dfrac{VCD^2}{DC^4V} =$

B. $\dfrac{\sqrt{FYU}}{F(YU)^3} =$

$$y = mx+b$$

C. $\dfrac{T^3}{\sqrt{T}} =$

When people have nothing else to do, they decide to invent Algebra and make some progress in our understanding of the world.

Solve for C:

A. $\dfrac{GHV}{A^2\sqrt{C}} = F$

B. $\dfrac{C^3B}{C} = C$

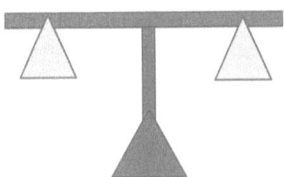

C. $\dfrac{C}{C^7} = D$

The main concept of Algebra is keeping a balance between numbers.

4. Convert the following numbers to Standard Scientific Notation and also state the number of significant figures:

A. $0.0009 =$

B. $890.4 =$

C. $6789.0 =$

D. $890 =$

E. $5678 =$

F. $78.0 =$

5. Solve for C:

A. $\dfrac{(9.0x10^3)(2.0x10^{-2})(1.0x10^1)}{(3.0x10^4)(1.0x10^{-8})} = C$

B. $\dfrac{(7.0 \times 10^2)(2.0 \times 10^{-1})(1.0 \times 10^{10})}{(14 \times 10^{12})(1.0 \times 10^{-18})} = C$

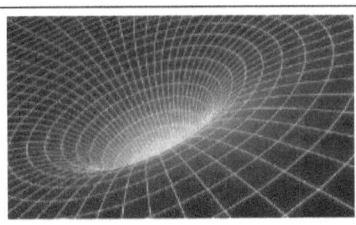

6. What are Newton's 3 Laws of Motion:

A=

B=

C=

7. How far will a ball drop in 2.00 s?

8. A 10.0 kg particle experiences a force of 30.0 N left and 50.0 N right.

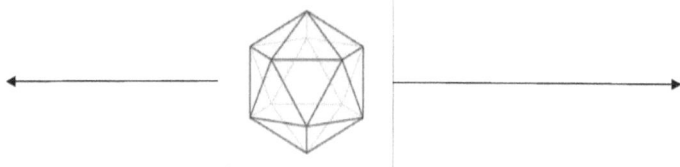

A. What is the Net Force with Magnitude and Direction?

B. What is the Acceleration?

C. What is the Displacement after 3.00 s?

10. How many times is the Circumference of a circle bigger than the Diameter?

Answer:_____

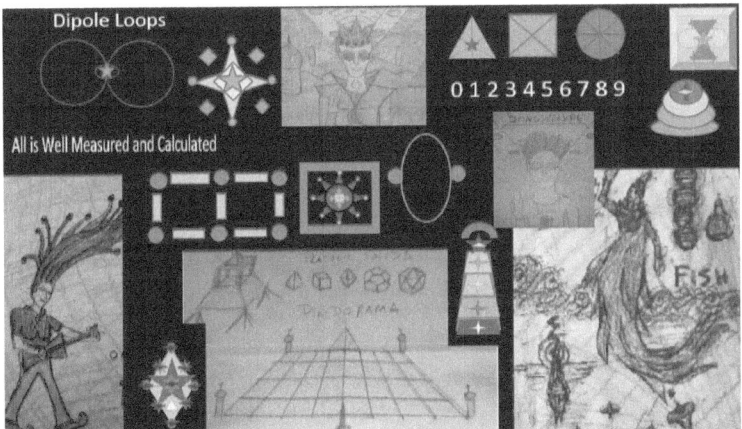

Work, Power, and Energy

The Work done on an object is the amount of Energy put on the object causing it to change speed or just move at constant speed. **Work= F(d)cos(angle between d and F)**

Where F is the Force applied and d is the Displacement of the object.

If Friction and Resistive Forces are ignored, the Work is equal to the change in Kinetic Energy of the object.

Work = ΔKE

Kinetic Energy is the Energy of motion. The faster something moves, the more Kinetic Energy that it has.

Kinetic Energy = $\frac{1}{2}mv^2$

Where m is the mass and v the speed of an object.

Potential Energy is stored Energy that can be released leading to motion.

According to the First Law of Thermodynamics the Total Energy of a Closed System is conserved. That means that Potential Energy becomes Kinetic and Kinetic becomes Potential.

Total Energy = Kinetic + Potential

Conservation	Gravitational Potential Energy is equal to mgh where m is the mass, g the acceleration due to gravity and h is the height above the surface of the Earth

Name_____ Period_____

Pulling a Box at an Angle 2

1) A 10.0 Kg box is pulled at an angle of 30° with a force of 150 N and $\mu = 0.12$. Find the following:

A) Weight:

B) Normal Force:

C) Force of Friction:

D) Net Force:

E) Acceleration:

F) What is the displacement after 1.00 minute?

G) What is the work done in 1.00 minute?

H) What is the Power delivered in that time?

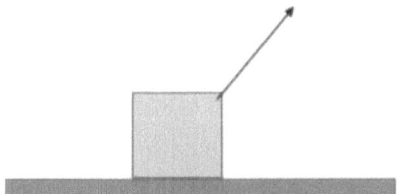

2) A 25.0 Kg box is pulled at an angle of 40° with a force of 234 N and $\mu = 0.16$. Find the following:

A) Weight:

B) Normal Force:

C) Force of Friction:

D) Net Force:

E) Acceleration:

F) What is the displacement after 1.00 minute?

G) What is the work done in 1.00 minute?

H) What is the Power delivered in that time?

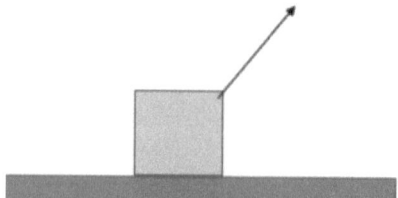

3) A 10.0 Kg box is pulled at an angle of 33° with a force of 99 N and $\mu = 0.20$. Find the following:

A) Weight:

B) Normal Force:

C) Force of Friction:

D) Net Force:

E) Acceleration:

F) What is the displacement after 30 seconds?

G) What is the Work done in 30 seconds?

H) What is the Power delivered in that time?

I) What is the Work done by Friction?

J) What is the Net Work done in 30 seconds?

K) Using Energy find the Final Velocity after 30 seconds:

4) A 80.0 Kg box is pulled at an angle of 43° with a force of 700 N and $\mu = 0.10$. Find the following:

A) Weight:

B) Normal Force:

C) Force of Friction:

D) Net Force:

E) Acceleration:

F) What is the displacement after 20 seconds?

G) What is the Work done in 20 seconds?

H) What is the Power delivered in that time?

I) What is the Work done by Friction?

J) What is the Net Work done in 20 seconds?

K) Using Energy find the Final Velocity after 20 seconds:

5) A 26.0 Kg box is pulled at an angle of $13°$ with a force of 300 N and $\mu = 0.01$. Find the following:

A) Weight:

B) Normal Force:

C) Force of Friction:

D) Net Force:

E) Acceleration:

F) What is the displacement after 20 seconds?

G) What is the Work done in 20 seconds?

H) What is the Power delivered in that time?

I) What is the Work done by Friction?

J) What is the Net Work done in 20 seconds?

K) Using Energy find the Final Velocity after 20 seconds:

Name_____ Period_____

Pushing a box at an Angle and doing Work

1....You push a 10 kg box with a force of 195 N, at an angle of 30° below the horizontal and $\mu = 0.09$. Answer the following:

A) Weight:

B) Normal Force:

C) Force of Friction:

D) Net Force:

E) Acceleration:

F) What is the displacement after 10 seconds?

G) What is the Work done in 10 seconds?

H) What is the Power delivered in that time?

I) What is the Work done by Friction?

J) What is the Net Work done in 10 seconds?

K) Using Energy find the Final Velocity after 10 seconds:

2... You push a 34 kg box with a force of 100 N, at an angle of $33°$ below the horizontal and $\mu = 0.11$. Answer the following:

A) Weight:

B) Normal Force:

C) Force of Friction:

D) Net Force:

E) Acceleration:

F) What is the displacement after 5 seconds?

G) What is the Work done in 5 seconds?

H) What is the Power delivered in that time?

I) What is the Work done by Friction?

J) What is the Net Work done in 5 seconds?

K) Using Energy find the Final Velocity after 5 seconds:

3...A 7.80 kg box is pushed with a Force of 760 N. The Coefficient of Friction is 0.8. Find the following:

A....Draw a Free Body Diagram for the Box:

B...What is the Acceleration of the box?

C...What is the time taken to cover a distance of 100m?

D...What is the Work done?

E...What is the Work done by Friction?

F...What is the Final Velocity after covering 100 m?

G...After reaching 100m, the force is removed and only Friction stays. How long will the box take to slow down to rest?

H...What is the Work done by Friction in the slow down process?

I...How much farther will the box move in the slow down process?

4...A 9.80 kg box is pushed with a Force of 990 N. The Coefficient of Friction is 0.6. Find the following:

A....Draw a Free Body Diagram for the Box:

B...What is the Acceleration of the box?

C...What is the time taken to cover a distance of 400m?

D...What is the Work done?

E...What is the Work done by Friction?

F...What is the Final Velocity after covering 400 m?

G...After reaching 400m, the force is removed and only Friction stays. How long will the box take to slow down to rest?

H...What is the Work done by Friction in the slow down process?

I...How much farther will the box move in the slow down process?

Conservation of Energy

According to the First Law of Thermodynamics, *Energy cannot be created nor destroyed only transformed*. That means that Energy in the universe is conserved since the very beginning of history at the Big Bang all the way to modern times. The amount of Energy is always kept the same but can be converted from Potential to Kinetic and Kinetic to Potential or could be transferred to generate Heat and Sound. Nothing is truly lost and all Energy is conserved in the cosmic history.

Total Energy = Kinetic + Potential + Dissipative

Kinetic Energy is Energy of motion and its equation is $(1/2)mv^2$

Gravitational Potential Energy for objects near the surface is mgh

M is the mass in kg, g is the acceleration due to gravity, h the height from surface, and v is the speed. For gravity on Earth's Surface the equation below works:

Total Energy = $(1/2)mv^2$ + mgh   (Ignoring Friction)

> The Equation above only works if Air Resistance is also ignored

Equations: **Energy is conserved ignoring Friction**

Energy for an Oscillating Pendulum:

Total Energy = mgh + $(1/2)mv^2$

Energy for an Oscillating Spring Mass System:

Total Energy = $(1/2)k\Delta x^2 + (1/2)mv^2$

The Restoring Force of a Spring is the Force that attempts to move the Spring back to Equilibrium Position:

Restoring Force = -kx

K is the Spring Constant which is the amount of Force needed to stretch or compress a spring 1.00 m.

Indicate the direction of the Restoring Force in each part of the oscillation below:

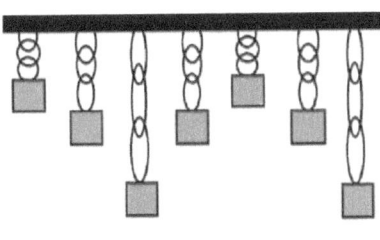

A=

B=

C=

D=

E=

F=

G=

Indicate the locations where the Restoring Force is zero:

Indicate the locations where the restoring Force is Max Negative:

Indicate the locations where the restoring Force is Max Positive:

Indicate the locations where Velocity is Max:

Indicate the locations where Velocity is zero:

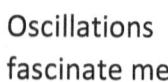

Indicate the locations where the Oscillation is at Equilibrium Position:

Indicate the direction of the Restoring Force in each part of the oscillation below:

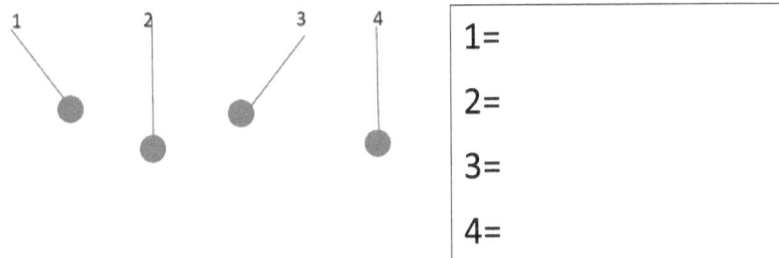

1=

2=

3=

4=

Indicate the locations where the Restoring Force is zero:

Indicate the locations where the restoring Force is Max Right:

Indicate the locations where the restoring Force is Max Left:

Indicate the locations where Velocity is Max:

Indicate the locations where Velocity is zero:

Indicate the locations where the Oscillation is at Equilibrium Position:

Solve the bottom problems using Energy:

1...A ball drops 10.0 m. What is its Final Velocity ignoring Air Resistance?

2...A ball is launched straight up at 5.00 m/s. What is the Maximum Height?

3...A ball is thrown straight down from the top of a building at -8.00 m/s. Its Final Velocity is -50.0m/s, how high is the building?

4.... A ball is thrown straight down from the top of a 300 m high building at -3.00 m/s. What is its Final Velocity?

Conservation of Energy Review

5... In the roller coaster below friction is considered negligible

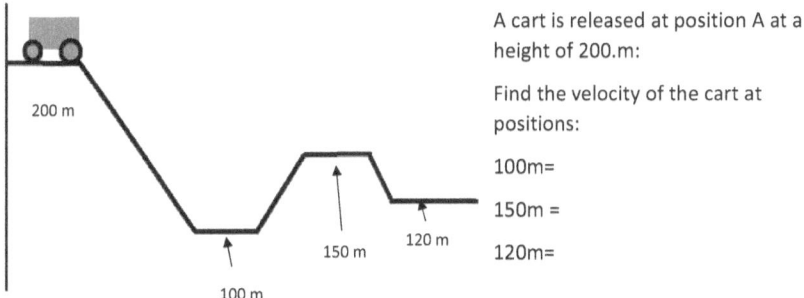

A cart is released at position A at a height of 200.m:

Find the velocity of the cart at positions:

100m=

150m =

120m=

6. If the cart in problem 5 has a mass of 30.0 kg what is the Total Energy of the System?

7. What is the First Law of Thermodynamics?

All Energy is wave. According to String Theory different vibration patterns lead to different particles.

8. In the roller coaster below friction is considered negligible

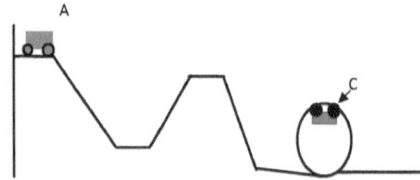

A cart is released at position A at a height of 25.0 m:

Find the velocity of the cart at position C, 10 m above the ground: _____

9. In the roller coaster below friction is considered negligible

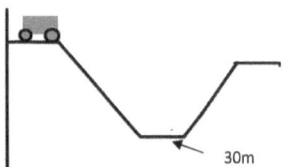

A cart is released at position A at a height of 70.0 m:

Find the velocity of the cart at height 30.0 m : _____

10. Measuring the Energy of a 40.0 kg Ball falling from a 3000 m high Tower:

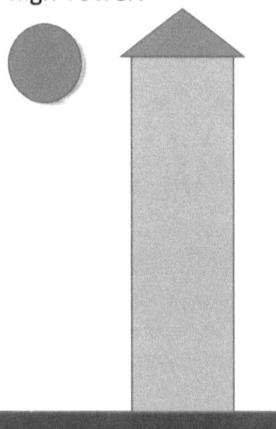

A. What is the Total Energy of the Ball?

_____

B. What is its Final Velocity after hitting the ground?

The ball is in free fall from the top of a Tower. Its Displacement within each second increases with the square of the time. Wow!

C. What is the Potential Energy of the ball on the ground?

D. What is the Kinetic Energy of the ball on ground?

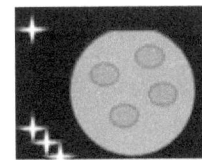

G. The ball hits the ground and bounces back up with 50% of its Initial Energy. How high does it rise?

E. What is the Kinetic Energy of the ball at the top of the Tower?

F. What is the Work done by Gravity throughout the fall?

Could it be that going through a wormhole is the same a free falling?

_____

Show your work in this box:

In the graph to the right please graph the Displacement Vs Time and the Velocity vs Time Graph for the ball falling from the Tower.

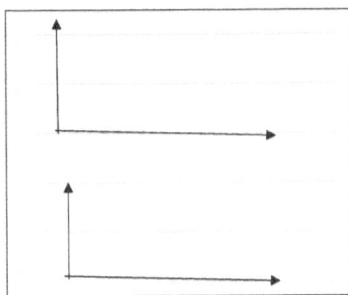

11...Zirvina is pulling a cart whose mass is 70.0 kg at an angle of 20° with a Force of 700 N: (Assume there is no Friction but the wheels do spin somehow).

A. What is the Work that Zirvina does on the cart by pulling it a distance of 800m?

B. What is the gain in Kinetic Energy of the cart?

C. What is the Final Velocity of the cart?

Please help Zirvina graph the Displacement and Velocity vs Time Graph for his work:

D. What is the Net Force on the cart?

E. What is the Acceleration of the cart?

F. How long does it take the cart to reach the distance of 800.m?

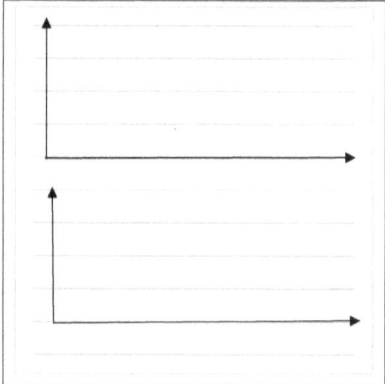

8. The MilkyBar Chocolate in the picture to the right has 175,000J of Energy. How many of these does Zirvina need to eat to recover from his hard work?

12....Label the Points of Maximum Potential Energy and Maximum Kinetic Energy:

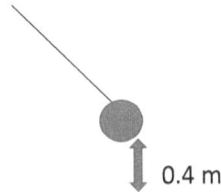 0.4 m

A. With what Velocity will the Pendulum be moving at the very bottom, or equilibrium position? It was dropped from a height of 0.4m:

13....Label the Points of Maximum Potential Energy and Maximum Kinetic Energy:

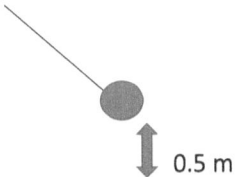 0.5 m

A. With what Velocity will the Pendulum be moving at the very bottom, or equilibrium position? It was dropped from a height of 0.5m:

14. Vasco da Gama arrived off the coast of South America and he is trying to help two locals discover how high they could jump if all the energy they acquired from drinking the water of coconut were to be converted to Kinetic Energy. The Coconut Water has 2,000 J. The mass of each of them is 50.0 kg.

Show your work here:

How high will they be able to jump?

9. Vasco da Gama decides to rest a little bit under the shade of a tree. A coconut then falls on his head. The velocity of the coconut was 9.00 m/s.

How high was the fall?

_____

Label the locations in the Oscillations:

15...State where is in the Oscillation of a Spring is the Kinetic and Potential Energies maximum or zero:

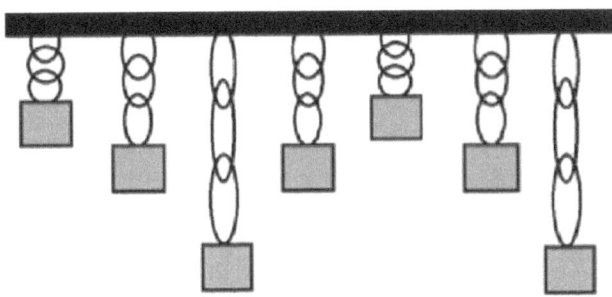

Max Potential Energy:

Max Kinetic Energy:

Zero Potential Energy:

Zero Kinetic Energy:

16…State where in the Oscillation of a Pendulum is the Kinetic and Potential Energies maximum or zero:

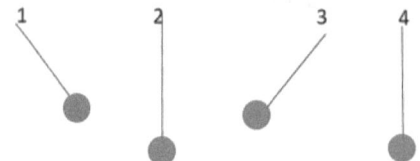

Max Potential Energy:

Max Kinetic Energy:

Zero Potential Energy:

Zero Kinetic Energy:

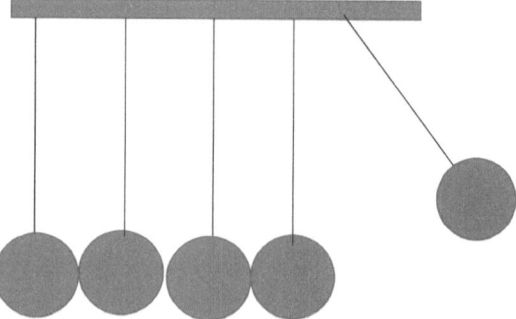

Waves 1 Review

17…For the Pendulum answer the following questions:

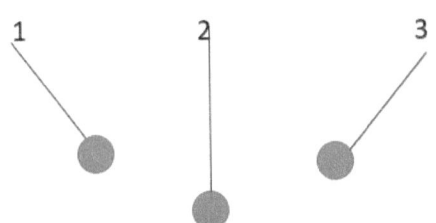
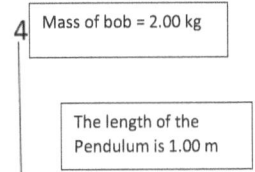

Mass of bob = 2.00 kg

The length of the Pendulum is 1.00 m

1…Where in the cycle is:

Kinetic Energy Max:_____

Potential Energy Max:_____

Kinetic Energy zero:_____

Potential Energy zero:_____

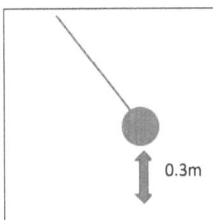

0.3m

2…If the Pendulum was dropped when the Bob was at a height of 0.3 m from its Equilibrium, how fast will it be moving when Kinetic Energy is Max?

What is the Total Energy of this oscillating Pendulum?_____

3..How fast will it be moving at Max Potential Energy? _____

4.If after each cycle the Pendulum loses 20% of its Energy, how fast will it be moving at Equilibrium Position after 1 cycle?

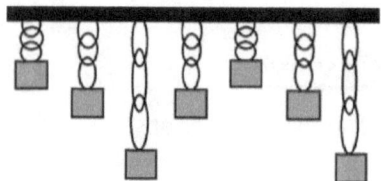

For the Spring Mass System of the left answer the following questions:

If the spring has K= 8.00 N/m, the hanging mass is 2.80 kg, and is Displaced 0.8 m:

1...Total energy of the oscillation:

2...State the locations of:

Max Potential Energy:_____

3...Where is the Max speed is reached?

Max Kinetic Energy:_____

Zero Potential Energy:_____

4...What is the Max speed?

Zero Kinetic Energy:_____

5...Where is the Speed zero?

6...If the Spring Mass System loses 20% of its energy is each oscillation, what is the Max Speed after 3 cycles?

7...How much will it compress and stretch on the third cycle?

18...Our little Tupinamba wants to know that if he throws his coconut ball up in the air with a Force of 400 N after accelerating it for 2.00s, and it reaches a height of 90.0m, what is the mass of the coconut ball?

$F\Delta t = m\Delta v$

$(1/2)mv^2 = mgh$

$F\Delta t = mv$

$\dfrac{F\Delta t}{v} = m$

Conservation of Energy Review 2

19. In the roller coaster below friction is considered negligible

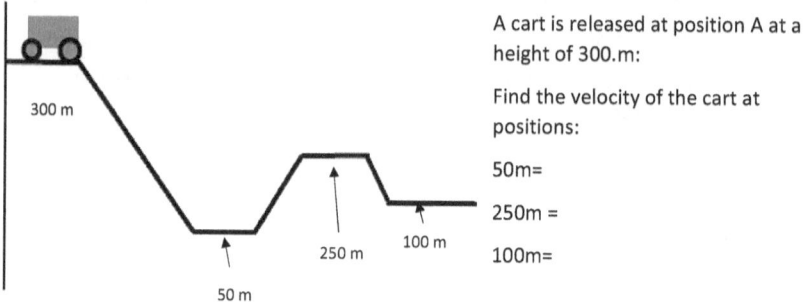

A cart is released at position A at a height of 300.m:

Find the velocity of the cart at positions:

50m=

250m =

100m=

If the cart in problem 19 has a mass of 70.0 kg what is the Total Energy of the System?

What is the First Law of Thermodynamics?

20. In the roller coaster below friction is considered negligible

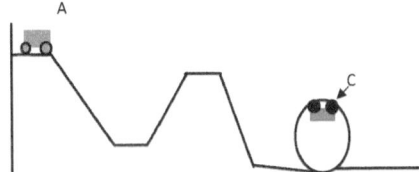

A cart is released at position A at a height of 55.0 m:

Find the velocity of the cart at position C, 40 m above the ground: _____

21. In the roller coaster below friction is considered negligible

A cart is released at position A at a height of 90.0 m:

Find the velocity of the cart at height 20.0 m : _____

22. Measuring the Energy of a 90.0 kg Ball falling from an 8000 m high Tower:

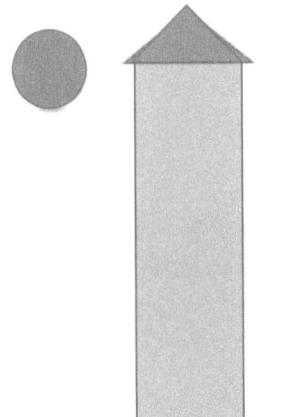

A. What is the Total Energy of the Ball?

_____

B. What is its Final Velocity after hitting the ground?

The ball is in free fall from the top of a Tower. Its Displacement within each second increases with the square of the time. Wow!

23. Zirvina is pulling a cart whose mass is 80.0 kg at an angle of 30° with a Force of 600 N: (Assume there is no Friction but the wheels do spin somehow).

A. What is the Work that Zirvina does on the cart by pulling it a distance of 400m?

B. What is the gain in Kinetic Energy of the cart?

C. What is the Final Velocity of the cart?

Please help Zirvina graph the Displacement and Velocity vs Time Graph for his work:

D. What is the Net Force on the cart?

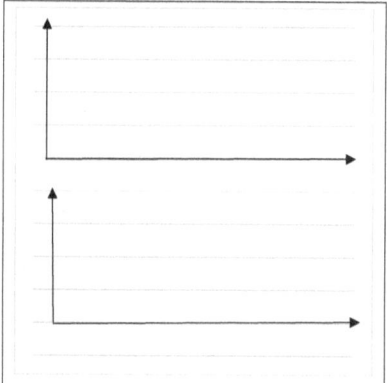

E. What is the Acceleration of the cart?

F. How long does it take the cart to reach the distance of 400.m?

8. The MilkyBar Chocolate in the picture to the right has 623538J of Energy. How many of these does Zirvina need to eat to recover from his hard work?

Milky Bar Chocolate

24...A ball thrown upward reaches a height of 10 m. How fast was it thrown?

25...A rock is thrown out of the top of a building that is 100 m high with a velocity of 6.00 m/s. Find the velocity of the ball when hitting the ground if the ball is:

Thrown horizontally:

Thrown Straight up:

Thrown straight down:

24...A person applies a variable force as shown in the graph below. Answer the following:

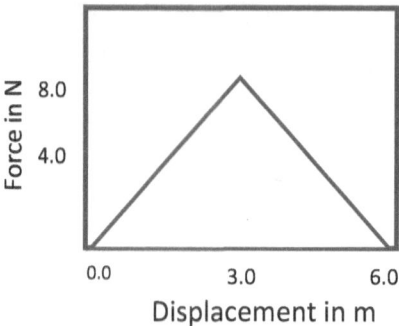

A...What is the Work done on the box?

B...If the box is 7.8 kg what is its gain in Kinetic Energy if there is no Friction?

C...What is the Final Velocity?

D...What is the Average Acceleration if this Work happened in 18 s?

E...What is the Power delivered to the box?

25...A person applies a variable force as shown in the graph below. Answer the following:

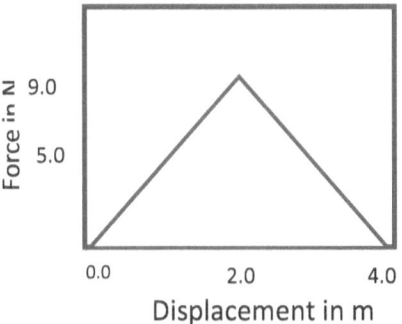

Displacement in m

A...What is the Work done on the box?

B...If the box is 1.8 kg what is its gain in Kinetic Energy if there is no Friction?

C...What is the Final Velocity?

D...What is the Average Acceleration if this Work happened in 22 s?

E...What is the Power delivered to the box?

26...A person applies a variable force as shown in the graph below. Answer the following:

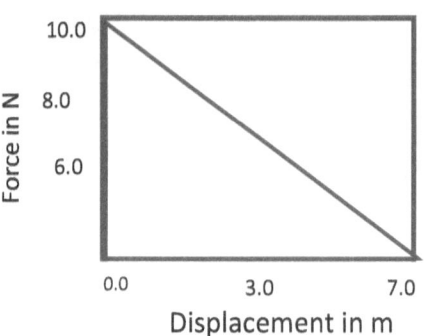

A...What is the Work done on the box?

B...If the box is 6.0 kg what is its gain in Kinetic Energy if there is no Friction?

C...What is the Final Velocity?

D...What is the Average Acceleration if this Work happened in 30 s?

E...What is the Power delivered to the box?

27...A person applies a variable force as shown in the graph below. Answer the following:

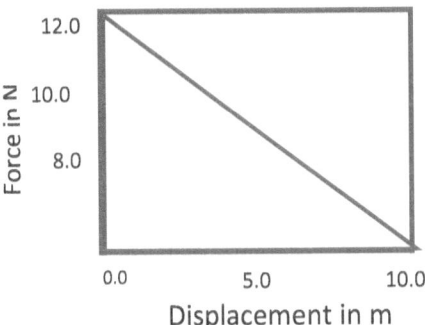

A...What is the Work done on the box?

B...If the box is 9.0 kg what is its gain in Kinetic Energy if there is no Friction?

C...What is the Final Velocity?

D...What is the Average Acceleration if this Work happened in 40 s?

E...What is the Power delivered to the box?

28...A person applies a variable force as shown in the graph below. Answer the following:

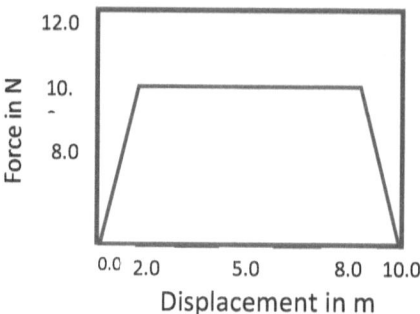

A...What is the Work done on the box?

B...If the box is 12.0 kg what is its gain in Kinetic Energy if there is no Friction?

C...What is the Final Velocity?

D...What is the Average Acceleration if this Work happened in 20 s?

E...What is the Power delivered to the box?

29...A person applies a variable force as shown in the graph below. Answer the following:

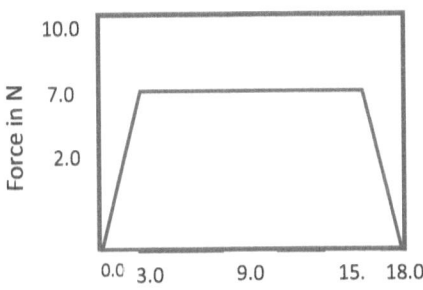

Displacement in m

A...What is the Work done on the box?

B...If the box is 12.0 kg what is its gain in Kinetic Energy if there is no Friction?

C...What is the Final Velocity?

D...What is the Average Acceleration if this Work happened in 20 s?

E...What is the Power delivered to the box?

Name_____ Period_____

Work Power Energy Review 1

1....A 20.0 Kg box is pulled at an angle of 30° with a force of 89 N and $\mu = 0.10$. Find the following:

A) Weight:

B) Normal Force:

C) Force of Friction:

D) Net Force:

E) Acceleration:

F) What is the displacement after 30 seconds?

G) What is the Work done in 30 seconds?

H) What is the Power delivered in that time?

I) What is the Work done by Friction?

J) What is the Net Work done in 30 seconds?

K) Using Energy find the Final Velocity after 30 seconds:

2....You push a 20 kg box with a force of 200 N, at an angle of $20°$ below the horizontal and $\mu = 0.15$. Answer the following:

A) Weight:

B) Normal Force:

C) Force of Friction:

D) Net Force:

E) Acceleration:

F) What is the displacement after 10 seconds?

G) What is the Work done in 10 seconds?

H) What is the Power delivered in that time?

I) What is the Work done by Friction?

J) What is the Net Work done in 10 seconds?

K) Using Energy find the Final Velocity after 10 seconds:

3...A 5.80 kg box is pushed with a Force of 800 N. The Coefficient of Friction is 0.3. Find the following:

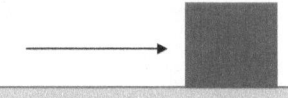

A....Draw a Free Body Diagram for the Box:

B...What is Friction?

C...What is the Net Force?

D...What is the Acceleration of the box?

C...What is the time taken to cover a distance of 300m?

D...What is the Work done?

E...What is the Work done by Friction?

F...What is the Final Velocity after covering 300 m?

G...After reaching 300m, the force is removed and only Friction stays. How long will the box take to slow down to rest?

H...What is the Work done by Friction in the slow down process?

I...How much farther will the box move in the slow down process?

4...A person applies a variable force as shown in the graph below. Answer the following:

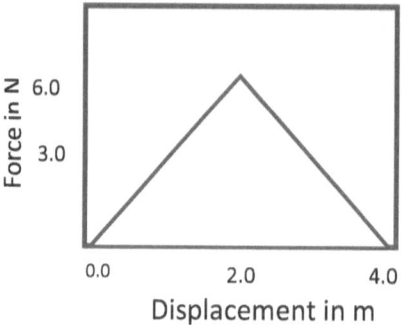

A...What is the Work done on the box?

B...If the box is 3.8 kg what is its gain in Kinetic Energy if there is no Friction?

C...What is the Final Velocity?

D...What is the Average Acceleration if this Work happened in 28 s?

E...What is the Power delivered to the box?

29...A person applies a variable force as shown in the graph below. Answer the following:

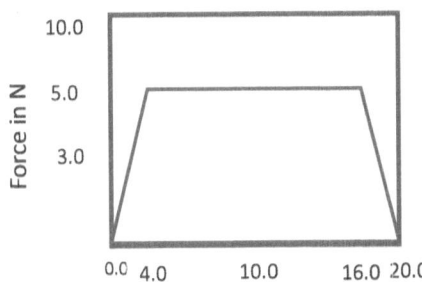

Displacement in m

A...What is the Work done on the box?

B...If the box is 17.0 kg what is its gain in Kinetic Energy if there is no Friction?

C...What is the Final Velocity?

D...What is the Average Acceleration if this Work happened in 30 s?

E...What is the Power delivered to the box?

Name_____Period_____

Work Power Energy Review 2

1....A 30.0 Kg box is pulled at an angle of 20° with a force of 210 N and $\mu = 0.20$. Find the following:

A) Weight:

B) Normal Force:

C) Force of Friction:

D) Net Force:

E) Acceleration:

F) What is the displacement after 10 seconds?

G) What is the Work done in 10 seconds?

H) What is the Power delivered in that time?

I) What is the Work done by Friction?

J) What is the Net Work done in 10 seconds?

K) Using Energy find the Final Velocity after 10 seconds:

2....You push a 10 kg box with a force of 200 N, at an angle of 40° below the horizontal and $\mu = 0.14$. Answer the following:

A) Weight:

B) Normal Force:

C) Force of Friction:

D) Net Force:

E) Acceleration:

F) What is the displacement after 2.00 seconds?

G) What is the Work done in 2.00 seconds?

H) What is the Power delivered in that time?

I) What is the Work done by Friction?

J) What is the Net Work done in 2.00 seconds?

K) Using Energy find the Final Velocity after 2.00 seconds:

3...A 1.80 kg box is pushed with a Force of 100 N. The Coefficient of Friction is 0.9. Find the following:

A.....Draw a Free Body Diagram for the Box:

B...What is Friction?

C...What is the Net Force?

D...What is the Acceleration of the box?

E...What is the time taken to cover a distance of 30.0m?

F...What is the Work done?

G...What is the Work done by Friction?

H...What is the Final Velocity after covering 30.0 m?

I...After reaching 30.0m, the force is removed and only Friction stays. How long will the box take to slow down to rest?

J...What is the Work done by Friction in the slow down process?

K...How much farther will the box move in the slow down process?

4...A person applies a variable force as shown in the graph below. Answer the following:

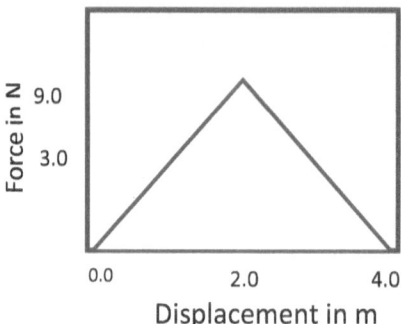

A...What is the Work done on the box?

B...If the box is 6.8 kg what is its gain in Kinetic Energy if there is no Friction?

C...What is the Final Velocity?

D...What is the Average Acceleration if this Work happened in 8.00 s?

E...What is the Power delivered to the box?

29...A person applies a variable force as shown in the graph below. Answer the following:

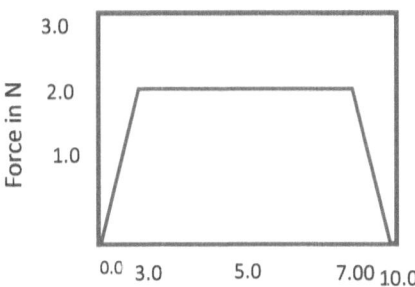

Displacement in m

A...What is the Work done on the box?

B...If the box is 7.0 kg what is its gain in Kinetic Energy if there is no Friction?

C...What is the Final Velocity?

D...What is the Average Acceleration if this Work happened in 3.0 s?

E...What is the Power delivered to the box?

Name_____ Period_____

Work Power Energy Test

1....A 25.0 Kg box is pulled at an angle of 21° with a force of 200 N and $\mu = 0.10$. Find the following:

A) Weight:

B) Normal Force:

C) Force of Friction:

D) Net Force:

E) Acceleration:

F) What is the displacement after 3.0 seconds?

G) What is the Work done in 3.0 seconds?

H) What is the Power delivered in that time?

I) What is the Work done by Friction?

J) What is the Net Work done in 3.0 seconds?

K) Using Energy find the Final Velocity after 3.0 seconds:

2....You push a 19 kg box with a force of 100 N, at an angle of $30°$ below the horizontal and $\mu = 0.10$. Answer the following:

A) Weight:

B) Normal Force:

C) Force of Friction:

D) Net Force:

E) Acceleration:

F) What is the displacement after 2.00 seconds?

G) What is the Work done in 2.00 seconds?

H) What is the Power delivered in that time?

I) What is the Work done by Friction?

J) What is the Net Work done in 2.00 seconds?

K) Using Energy find the Final Velocity after 2.00 seconds:

3...A 2.89 kg box is pushed with a Force of 30 N. The Coefficient of Friction is 0.9. Find the following:

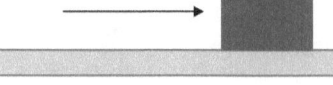

A....Draw a Free Body Diagram for the Box:

B...What is Friction?

C...What is the Net Force?

D...What is the Acceleration of the box?

C...What is the time taken to cover a distance of 10.0m?

D...What is the Work done?

E...What is the Work done by Friction?

F...What is the Final Velocity after covering 10.0 m?

G...After reaching 10.0m, the force is removed and only Friction stays. How long will the box take to slow down to rest?

H...What is the Work done by Friction in the slow down process?

I...How much farther will the box move in the slow down process?

4...A person applies a variable force as shown in the graph below. Answer the following:

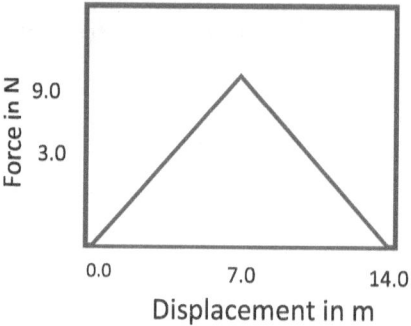

A...What is the Work done on the box?

B...If the box is 7.8 kg what is its gain in Kinetic Energy if there is no Friction?

C...What is the Final Velocity?

D...What is the Average Acceleration if this Work happened in 3.00 s?

E...What is the Power delivered to the box?

29...A person applies a variable force as shown in the graph below. Answer the following:

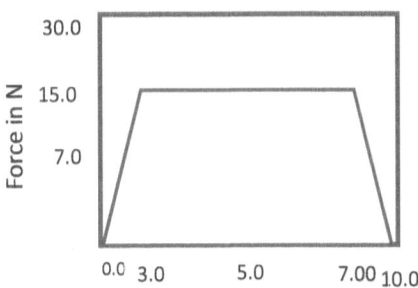

Displacement in m

A...What is the Work done on the box?

B...If the box is 9.0 kg what is its gain in Kinetic Energy if there is no Friction?

C...What is the Final Velocity?

D...What is the Average Acceleration if this Work happened in 1.0 s?

E...What is the Power delivered to the box?

Solve the bottom problems using Energy:

1...A ball drops 20.0 m. What is its Final Velocity ignoring Air Resistance?

2...A ball is launched straight up at 9.00 m/s. What is the Maximum Height?

3...A ball is thrown straight down from the top of a building at -1.00 m/s. Its Final Velocity is -80.0m/s, how high is the building?

4.... A ball is thrown straight down from the top of a 300 m high building at -4.00 m/s. What is its Final Velocity?

Conservation of Energy Review

5... In the roller coaster below friction is considered negligible

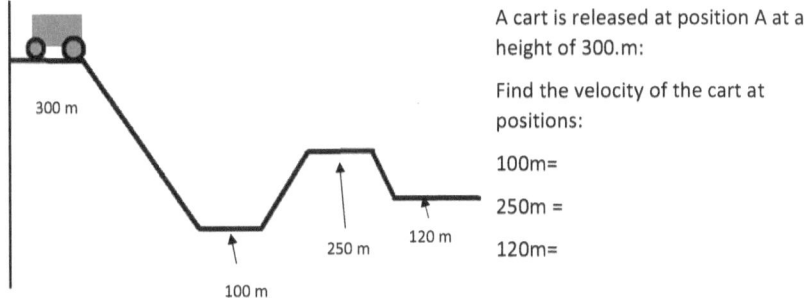

A cart is released at position A at a height of 300.m:

Find the velocity of the cart at positions:

100m=

250m =

120m=

6.If the cart in problem 5 has a mass of 40.0 kg what is the Total Energy of the System?

7.What is the First Law of Thermodynamics?

8. In the roller coaster below friction is considered negligible

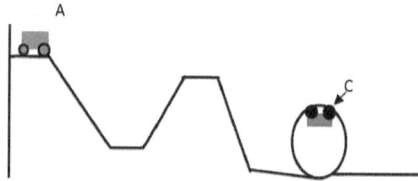

A cart is released at position A at a height of 35.0 m:

Find the velocity of the cart at position C, 10 m above the ground: _____

9. In the roller coaster below friction is considered negligible

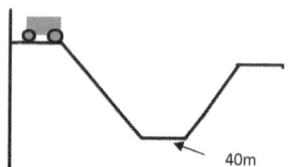

A cart is released at position A at a height of 80.0 m:

Find the velocity of the cart at height 40.0 m : _____

10. Measuring the Energy of a 50.0 kg Ball falling from a 7000 m high Tower:

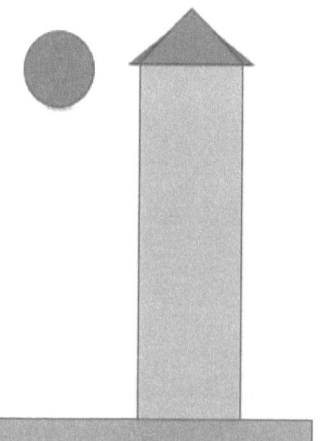

A. What is the Total Energy of the Ball?

_____

B. What is its Final Velocity after hitting the ground?

The ball is in free fall from the top of a Tower. Its Displacement within each second increases with the square of the time. Wow!

C. What is the Potential Energy of the ball on the ground?

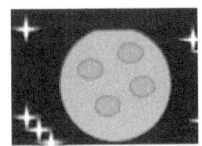

D. What is the Kinetic Energy of the ball on ground?

G. The ball hits the ground and bounces back up with 30% of its Initial Energy. How high does it rise?

E. What is the Kinetic Energy of the ball at the top of the Tower?

Could it be that going through a wormhole is the same a free falling?

F. What is the Work done by Gravity throughout the fall?

_____

Show your work in this box:

In the graph to the right please graph the Displacement Vs Time and the Velocity vs Time Graph for the ball falling from the Tower.

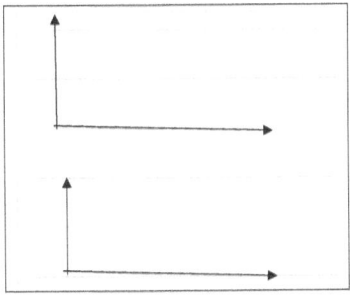

11...Zirvina is pulling a cart whose mass is 80.0 kg at an angle of 20° with a Force of 800 N: (Assume there is no Friction but the wheels do spin somehow).

A. What is the Work that Zirvina does on the cart by pulling it a distance of 900m?

B. What is the gain in Kinetic Energy of the cart?

C. What is the Final Velocity of the cart?

Please help Zirvina graph the Displacement and Velocity vs Time Graph for his work:

D. What is the Net Force on the cart?

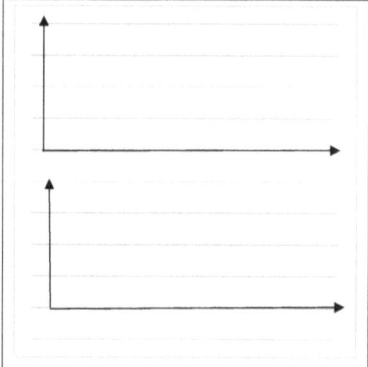

E. What is the Acceleration of the cart?

F. How long does it take the cart to reach the distance of 900.m?

8. The MilkyBar Chocolate in the picture to the right has 135,000 of Energy. How many of these does Zirvina need to eat to recover from his hard work?

Milky Bar Chocolate

12....Label the Points of Maximum Potential Energy and Maximum Kinetic Energy:

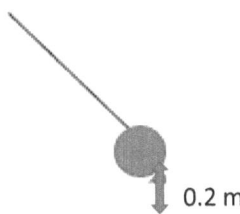

0.2 m

A. With what Velocity will the Pendulum be moving at the very bottom, or equilibrium position? It was dropped from a height of 0.2m:

13....Label the Points of Maximum Potential Energy and Maximum Kinetic Energy:

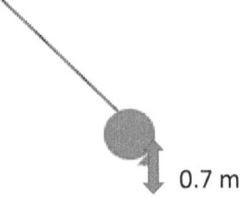

0.7 m

A. With what Velocity will the Pendulum be moving at the very bottom, or equilibrium position? It was dropped from a height of 0.7m:

14. Vasco da Gama arrived off the coast of South America and he is trying to help two locals discover how high they could jump if all the energy they acquired from drinking the water of coconut were to be converted to Kinetic Energy. The Coconut Water has 4,000 J. The mass of each of them is 30.0 kg.

Show your work here:

How high will they be able to jump?

9. Vasco da Gama decides to rest a little bit under the shade of a tree. A coconut then falls on his head. The velocity of the coconut was 11.0 m/s.

How high was the fall?

_____

Label the locations in the Oscillations:

15…State where is in the Oscillation of a Spring is the Kinetic and Potential Energies maximum or zero:

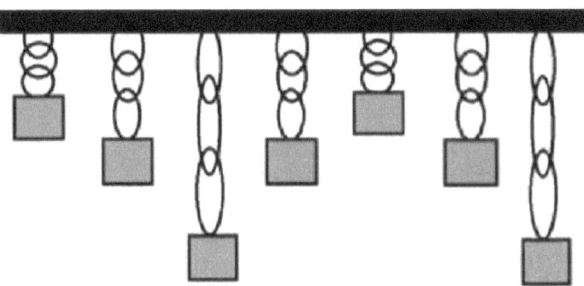

Max Potential Energy:

Max Kinetic Energy:

Zero Potential Energy:

Zero Kinetic Energy:

16...State where in the Oscillation of a Pendulum is the Kinetic and Potential Energies maximum or zero:

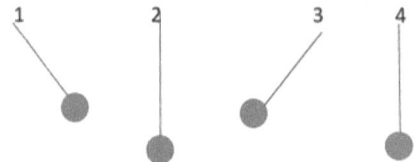

Max Potential Energy:

Max Kinetic Energy:

Zero Potential Energy:

Zero Kinetic Energy:

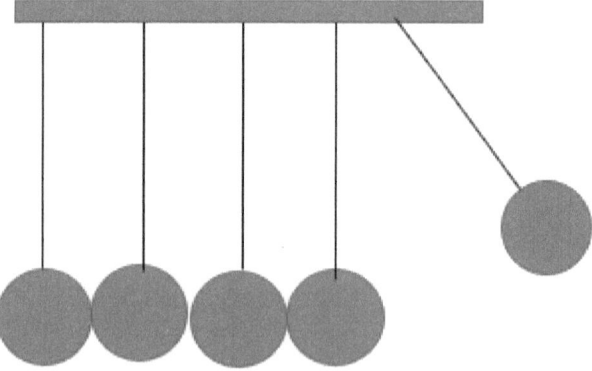

Waves 1 Review

17…For the Pendulum answer the following questions:

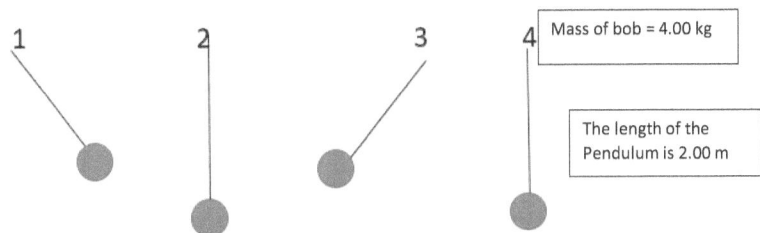

Mass of bob = 4.00 kg

The length of the Pendulum is 2.00 m

1…Where in the cycle is:

Kinetic Energy Max:_____

Potential Energy Max:_____

Kinetic Energy zero:_____

Potential Energy zero:_____

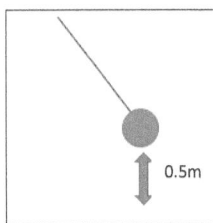

0.5m

2…If the Pendulum was dropped when the Bob was at a height of 0.5 m from its Equilibrium, how fast will it be moving when Kinetic Energy is Max?

What is the Total Energy of this oscillating Pendulum?_____

3..How fast will it be moving at Max Potential Energy? _____

4.If after each cycle the Pendulum loses 10% of its Energy, how fast will it be moving at Equilibrium Position after 2 cycles?

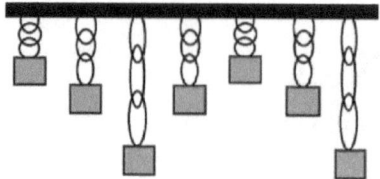

For the Spring Mass System of the left answer the following questions:

If the spring has K= 3.00 N/m, the hanging mass is 1.80 kg, and is Displaced 0.9 m:

1...Total energy of the oscillation:

2...State the locations of:

Max Potential Energy:_____

3...Where is the Max speed is reached?

Max Kinetic Energy:_____

Zero Potential Energy:_____

4...What is the Max speed?

Zero Kinetic Energy:_____

5...Where is the Speed zero?

6...If the Spring Mass System loses 10% of its energy is each oscillation, what is the Max Speed after 2 cycles?

7...How much will it compress and stretch on the third cycle?

Conservation of Energy Review 2

19. In the roller coaster below friction is considered negligible

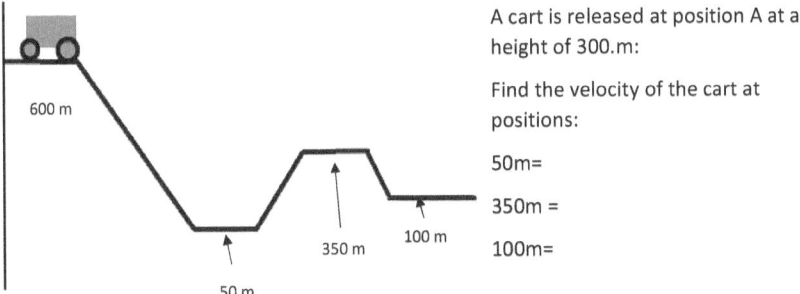

A cart is released at position A at a height of 300.m:

Find the velocity of the cart at positions:

50m=

350m =

100m=

If the cart in problem 1 has a mass of 20.0 kg what is the Total Energy of the System?

What is the First Law of Thermodynamics?

20. In the roller coaster below friction is considered negligible

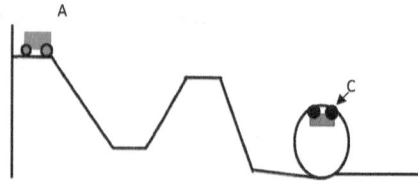

A cart is released at position A at a height of 65.0 m:

Find the velocity of the cart at position C, 20 m above the ground: _____

21. In the roller coaster below friction is considered negligible

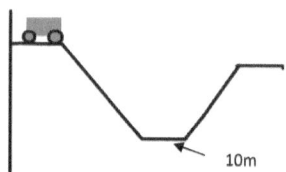

A cart is released at position A at a height of 80.0 m:

Find the velocity of the cart at height 10.0 m : _____

22. Measuring the Energy of a 90.0 kg Ball falling from a 1000 m high Tower:

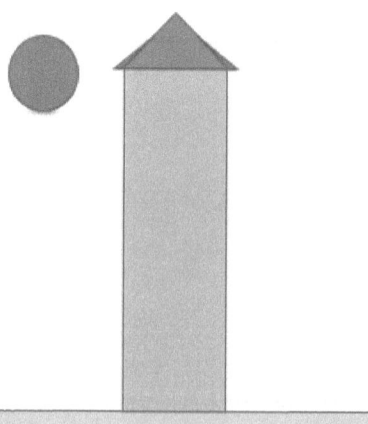

A. What is the Total Energy of the Ball?

_____

B. What is its Final Velocity after hitting the ground?

The ball is in free fall from the top of a Tower. Its Displacement within each second increases with the square of the time. Wow!

23. Zirvina is pulling a cart whose mass is 50.0 kg at an angle of 30° with a Force of 500 N: (Assume there is no Friction but the wheels do spin somehow).

A. What is the Work that Zirvina does on the cart by pulling it a distance of 400m?

B. What is the gain in Kinetic Energy of the cart?

C. What is the Final Velocity of the cart?

Please help Zirvina graph the Displacement and Velocity vs Time Graph for his work:

D. What is the Net Force on the cart?

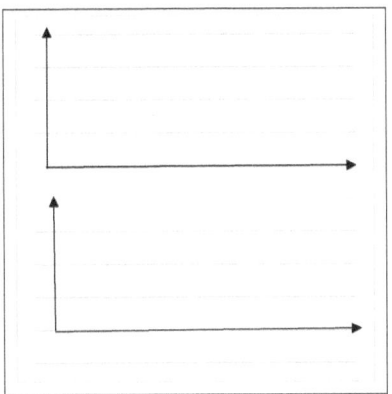

E. What is the Acceleration of the cart?

F. How long does it take the cart to reach the distance of 400.m?

8. The MilkyBar Chocolate in the picture to the right has 28,868J of Energy. How many of these does Zirvina need to eat to recover from his hard work?

Milky Bar Chocolate

24...A ball thrown upward reaches a height of 20 m. How fast was it thrown?

25...A rock is thrown out of the top of a building that is 200 m high with a velocity of 5.00 m/s. Find the velocity of the ball when hitting the ground if the ball is:

Thrown horizontally:

Thrown Straight up:

Thrown straight down:

Equations:	Pulling a Box:
Pendulum and Roller Coaster:	Work = $Fd\cos(\theta)$
Total Energy = $(1/2)mv^2 + mgh$	Net Force = $F\cos(\theta)$ – Friction
$V_f = \sqrt{2g\Delta h}$	
Spring Mass System:	Work = ΔKE
Total Energy = $(1/2)k\Delta x^2 + (1/2)mv^2$	KE = $(1/2)mv^2$
Restoring Force:	
F= -kx	

Name_____Period_____

Test Conservation of Energy

1...

Indicate the direction of the Restoring Force in each part of the oscillation below:

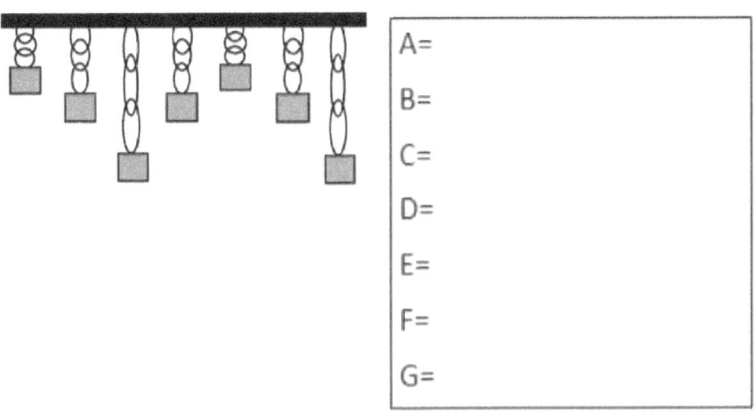

A=

B=

C=

D=

E=

F=

G=

Indicate the locations where the Restoring Force is zero:

Indicate the locations where the restoring Force is Max Negative:

Indicate the locations where the restoring Force is Max Positive:

Indicate the locations where Velocity is Max:

Indicate the locations where Velocity is zero:

Indicate the locations where the Oscillation is at Equilibrium Position:

Indicate the direction of the Restoring Force in each part of the oscillation below:

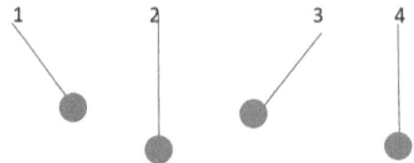

Indicate the locations where the Restoring Force is zero:

Indicate the locations where the restoring Force is Max Right:

Indicate the locations where the restoring Force is Max Left:

Indicate the locations where Velocity is Max:

Indicate the locations where Velocity is zero:

Indicate the locations where the Oscillation is at Equilibrium Position:

Solve the bottom problems using Energy:

2...A ball drops 30.0 m. What is its Final Velocity ignoring Air Resistance?

3...A ball is launched straight up at 8.00 m/s. What is the Maximum Height?

4...A ball is thrown straight down from the top of a building at -3.00 m/s. Its Final Velocity is -90.0 m/s, how high is the building?

5.... A ball is thrown straight down from the top of a 100 m high building at -3.00 m/s. What is its Final Velocity?

Conservation of Energy Review

6... In the roller coaster below friction is considered negligible

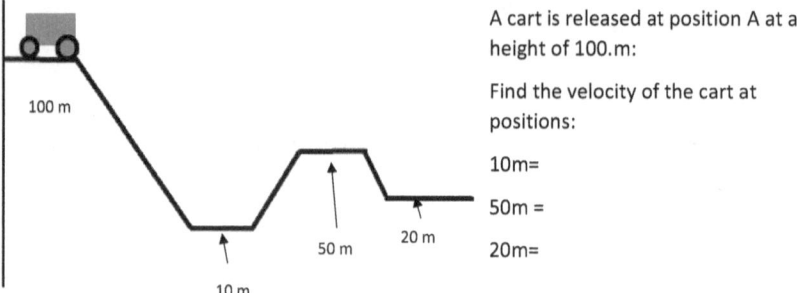

A cart is released at position A at a height of 100.m:

Find the velocity of the cart at positions:

10m=

50m =

20m=

If the cart in problem 6 has a mass of 20.0 kg what is the Total Energy of the System?

7. What is the First Law of Thermodynamics?

8. In the roller coaster below friction is considered negligible

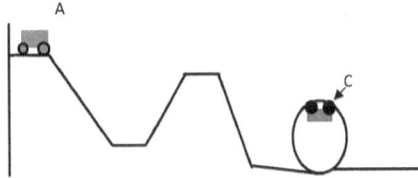

A cart is released at position A at a height of 45.0 m:

Find the velocity of the cart at position C, 10 m above the ground: _____

9. In the roller coaster below friction is considered negligible

A cart is released at position A at a height of 80.0 m:

Find the velocity of the cart at height 10.0 m : _____

10. Measuring the Energy of a 50.0 kg Ball falling from a 1000 m high Tower:

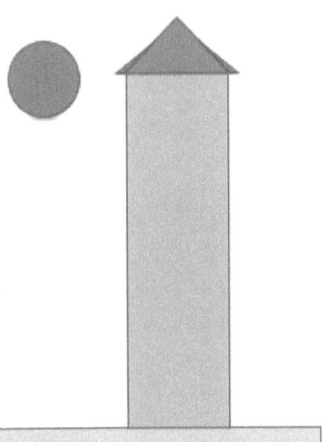

A. What is the Total Energy of the Ball?

_____

B. What is its Final Velocity after hitting the ground?

The ball is in free fall from the top of a Tower. Its Displacement within each second increases with the square of the time. Wow!

C. What is the Potential Energy of the ball on the ground?

D. What is the Kinetic Energy of the ball on ground?

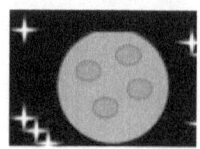

G. The ball hits the ground and bounces back up with 20% of its Initial Energy. How high does it rise?

E. What is the Kinetic Energy of the ball at the top of the Tower?

F. What is the Work done by Gravity throughout the fall?

Could it be that going through a wormhole is the same a free falling?

_____

11....Answer the following for the Pendulum below:

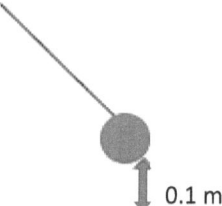

0.1 m

A. With what Velocity will the Pendulum be moving at the very bottom, or equilibrium position? It was dropped from a height of 0.1m:

Name_____Period_____

Momentum and Impulse

Momentum is a Vector quantity. Its equation is Mass times Velocity.

Momentum in a Closed System is conserved.

There are Two Types of Collisions:

Elastic Collision where both Momentum and Kinetic Energy is Conserved.

Inelastic Collision where only Momentum is conserved.

The area under a curve of a Graph of Force vs Displacement is the Work.

The area under a curve of a Graph of Force vs Time is Impulse.

Impulse is Change of Momentum.

Impulse = F$\Delta t = m\Delta v$

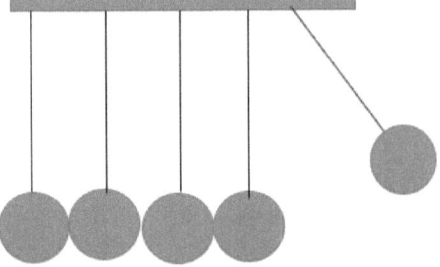

1...A 10.0 kg ball moves with a Velocity of 5.00 m/s to the right.

There are so many collisions between particles

A...What is its Momentum?

B...The ball collides with another ball of mass 10.0 kg and then stops. What is the Velocity of the second ball?

C...What is the change in Kinetic Energy?

Yurgue! I am more worried about calculating their momentum!

D... What kind of collision is it?

2...A 20.0 kg ball moves with a Velocity of 7.00 m/s to the right.

Zirvina! Particle collisions are also conceptually amazing!

A...What is its Momentum?

B...The ball collides with a 7.00 kg ball and they both stick together. What is the Velocity of both moving together?

C...What is the change in Kinetic Energy?

D... What kind of collision is it?

3...A 30.0 kg ball moves with a Velocity of 9.00 m/s to the right.

A...What is its Momentum?

B...The ball collides with a 20.0 kg ball and they both stick together. What is the Velocity of both moving together?

C...What is the change in Kinetic Energy?

D... What kind of collision is it?

Name_____Period____

Review Momentum and Impulse 3 SIG FIGS in this TEST

1...A Particle of mass 24.0 kg moves with Velocity 7.00 m/s to left. It then enters a Field in which it experiences a Force of 7.00 N to the right for 7.00 s.

The name of the Particle is LaydeZerink!

A...What is the Initial Momentum of the Particle?

B...What is the Impulse of the Particle?

C...What is the change in Momentum of the Particle?

D...What is the Final Velocity of the Particle?

E…What is the gain in Kinetic Energy of the Particle?

F…What was the Acceleration of the Particle in that time?

G…What was the Work done on the Particle?

H….What is the Power that the Force delivered to LaydeZerink in that time?

2…The same Particle then enters another Field in which it now only experiences a Force of 2.00 N to the right for 30.0s.

A…What is the Impulse of the particle?

B...What is the change in Momentum of the Particle?

C...What is the Final Velocity of the Particle?

D...What is the Kinetic Energy of the Particle at the end of this time?

E...What was the Acceleration of the Particle in that time?

F...What was the Work done on the Particle?

G...What is the Power delivered to the LaydeZerink in that time?

3...The same Particle then collides with another particle that is at rest. The other particle has a mass of 0.0 kg. They both stick together.

LaydeZerink!

Brashalonix is the name of the other Particle!

Initial Condition

I see a cosmic event in Andromeda!

The two formed the most powerful ATOM That is not made of Protons.

Final Condition

A...What is the Final Velocity of the two particles moving together?

B...What is the loss of Kinetic Energy after collision?

C...State 4 places where the Energy lost go:

1...   3....

2...   4....

D...Explain why this loss of Kinetic Energy does not violate the First Law of Thermodynamics:

_____
_____
_____

4...A 3,000 kg soccer ball is moving with a Velocity of 60 m/s to the right towards a wall. It hits the wall and experiences a Force depicted in the graph below:

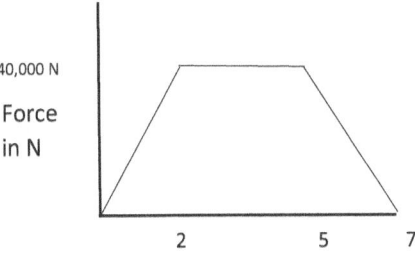

Time in s

A...What is the Impulse on the ball?

Please place a negative sign in front of the area under the curve:_____

B...What is the change in Momentum of the ball?

C...What is the Final Velocity of the ball? ____

Use: - (Area under curve) = m (Vf – 60)

5...A 3.00 kg soccer ball is dropped from a building that is 7000 m:

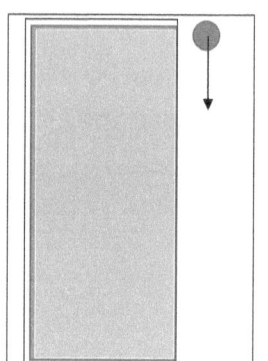

A... What is the Final Momentum of the ball when it hits the ground?

B... The ball hits the ground and comes to rest at 4.5 ms. What is the Force that the ground exerted on the ball?

> The strangest thing about the universe is that it is comprehensible.

BONUS

A...From a faraway planet, our character called Zirg is sent through a wormhole to the Earth. The wormhole pulls him with a Force of 10,000N for 30000s. What is his gain in Velocity?

_____

Mass of Zirg is 72 kg.

The speed of Light is 3×10^8 m/s. How many times is his speed bigger than light?

_____ (times)

Travel

What is the Impulse on a 32.0 kg object that is accelerated from 22.0 m/s to the speed of light in one hour?

What distance will the object have travelled in that time?

How many times is this distance larger than the separation between the Earth and the Sun?

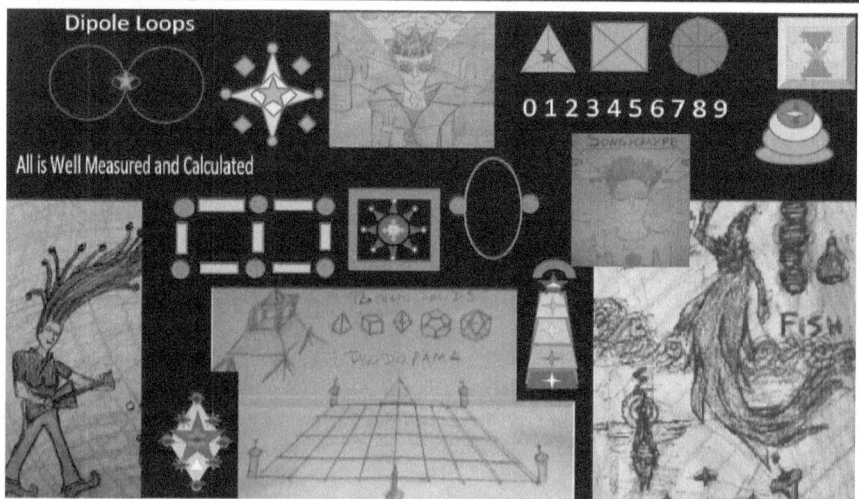

1...A person applies a variable force as shown in the graph below. Answer the following:

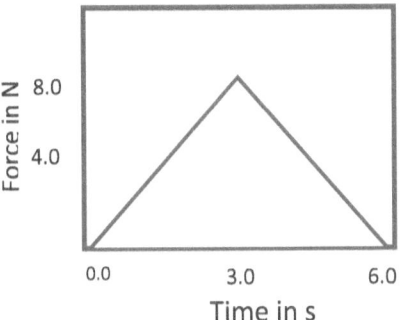

A...What is the Impulse on the box?

B...If the box is 7.8 kg what is its gain in Kinetic Energy if there is no Friction?

C...What is the Final Velocity?

D...What is the Average Acceleration?

E...What is the Power delivered to the box?

2...A person applies a variable force as shown in the graph below. Answer the following:

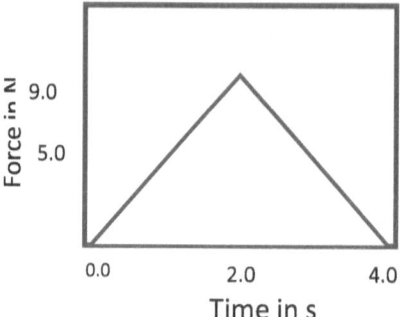

A...What is the Impulse on the box?

B...If the box is 9.8 kg what is its gain in Kinetic Energy if there is no Friction?

C...What is the Final Velocity?

D...What is the Average Acceleration?

E...What is the Power delivered to the box?

3...A person applies a variable force as shown in the graph below. Answer the following:

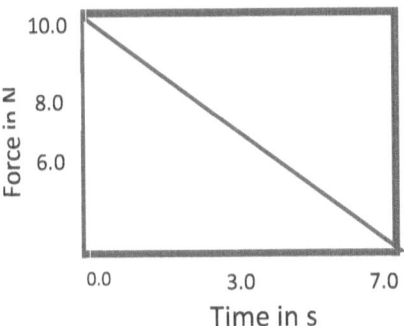

A...What is the Impulse on the box?

B...If the box is 1.2 kg what is its gain in Kinetic Energy if there is no Friction?

C...What is the Final Velocity?

D...What is the Average Acceleration?

E...What is the Power delivered to the box?

4...A person applies a variable force as shown in the graph below. Answer the following:

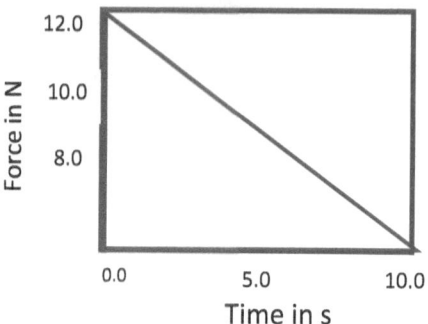

A...What is the Impulse on the box?

B...If the box is 8.0 kg what is its gain in Kinetic Energy if there is no Friction?

C...What is the Final Velocity?

D...What is the Average Acceleration?

E...What is the Power delivered to the box?

5...A person applies a variable force as shown in the graph below. Answer the following:

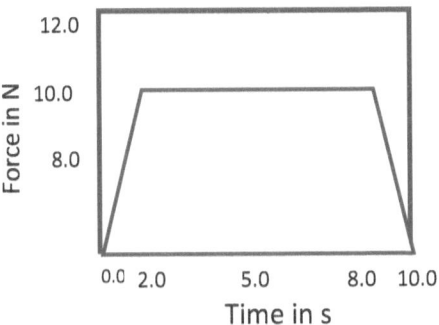

A...What is the Impulse on the box?

B...If the box is 0.8 kg what is its gain in Kinetic Energy if there is no Friction?

C...What is the Final Velocity?

D...What is the Average Acceleration?

E...What is the Power delivered to the box?

6...A person applies a variable force as shown in the graph below. Answer the following:

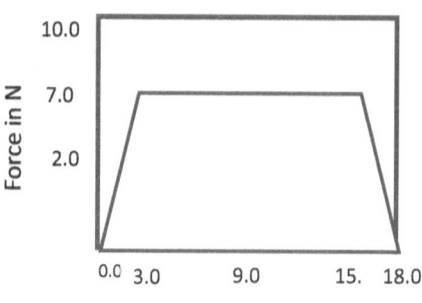

Time in s

A...What is the Impulse on the box?

B...If the box is 3.0 kg what is its gain in Kinetic Energy if there is no Friction?

C...What is the Final Velocity?

D...What is the Average Acceleration?

E...What is the Power delivered to the box?

Name_____Period_____

Momentum and Impulse 2

1...A 30.0 kg ball moves with a Velocity of 7.00 m/s to the right.

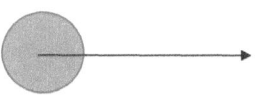

A...What is its Momentum?

B...The ball collides with another ball of mass 30.0 kg and then stops. What is the Velocity of the second ball?

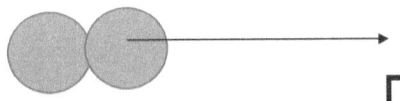

C...What is the change in Kinetic Energy?

D... What kind of collision is it?

2...A 30.0 kg ball moves with a Velocity of 9.00 m/s to the right.

Zirvina! Particle collisions are also conceptually amazing!

A...What is its Momentum?

B...The ball collides with a 9.00 kg ball and they both stick together. What is the Velocity of both moving together?

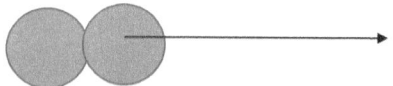

C...What is the change in Kinetic Energy?

D... What kind of collision is it?

3...A 50.0 kg ball moves with a Velocity of 3.00 m/s to the right.

A...What is its Momentum?

B...The ball collides with a 20.0 kg ball and they both stick together. What is the Velocity of both moving together?

C...What is the change in Kinetic Energy?

D... What kind of collision is it?

C...State 4 places where the Energy lost go:

1...   3....

2...   4....

D...Explain why this loss of Kinetic Energy does not violate the First Law of Thermodynamics:

_____
_____
_____

4...A 2,000 kg soccer ball is moving with a Velocity of 70 m/s to the right towards a wall. It hits the wall and experiences a Force depicted in the graph below:

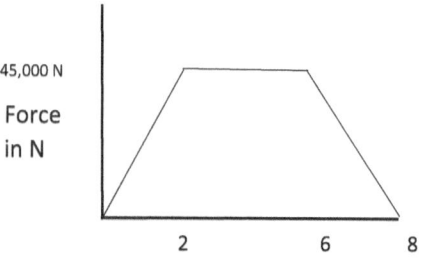

Time in s

A...What is the Impulse on the ball?

Please place a negative sign in front of the area under the curve: _____

B...What is the change in Momentum of the ball?

C...What is the Final Velocity of the ball? _____

Use: - (Area under curve) = m (Vf – 70)

5...A 6.00 kg soccer ball is dropped from a building that is 7000 m:

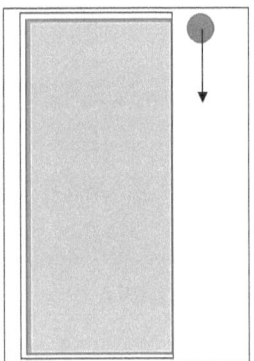

A... What is the Final Momentum of the ball when it hits the ground?

B... The ball hits the ground and comes to rest at 3.5 ms. What is the Force that the ground exerted on the ball?

The strangest thing about the universe is that it is comprehensible.

6...A 1,000 kg soccer ball is moving with a Velocity of 30 m/s to the right towards a wall. It hits the wall and experiences a Force depicted in the graph below:

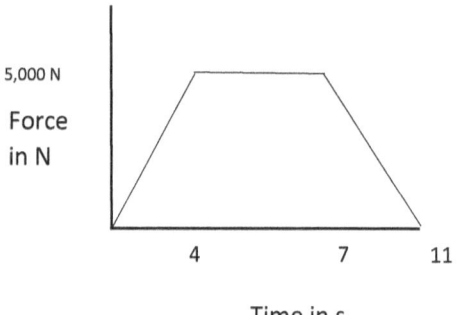

Time in s

A...What is the Impulse on the ball?

Please place a negative sign in front of the area under the curve: _____

B...What is the change in Momentum of the ball?

C...What is the Final Velocity of the ball? _____

Use: - (Area under curve) = m (Vf – 30)

Name_____ Period_____

Momentum and Impulse 3

1...A Particle of mass 34.0 kg moves with Velocity 9.00 m/s to left. It then enters a Field in which it experiences a Force of 24.0 N to the right for 12.75 s.

> The name of the Particle is LaydeZerink!

A...What is the Initial Momentum of the Particle?

B...What is the Impulse of the Particle?

C...What is the change in Momentum of the Particle?

D...What is the Final Velocity of the Particle?

E...What is the gain in Kinetic Energy of the Particle?

F...What was the Acceleration of the Particle in that time?

G...What was the Work done on the Particle?

H....What is the Power that the Force delivered to LaydeZerink in that time?

2...The same Particle then enters another Field in which it now only experiences a Force of 9.00 N to the right for 20.0s.

A...What is the Impulse of the particle?

B...What is the change in Momentum of the Particle?

C...What is the Final Velocity of the Particle?

D...What is the Kinetic Energy of the Particle at the end of this time?

E...What was the Acceleration of the Particle in that time?

F...What was the Work done on the Particle?

G...What is the Power delivered to the LaydeZerink in that time?

3...The same Particle then collides with another particle that is at rest. The other particle has a mass of 40.0 kg. They both stick together.

Initial Condition

Brashalonix is the name of the other Particle!

I see a cosmic event in Andromeda!

The two formed the most powerful ATOM That is not made of Protons.

Final Condition

A...What is the Final Velocity of the two particles moving together?

B...What is the loss of Kinetic Energy after collision?

4...For the system below the hanging mass is 4.00 kg and the mass over the frictionless table is 5.00 kg. Find the following:

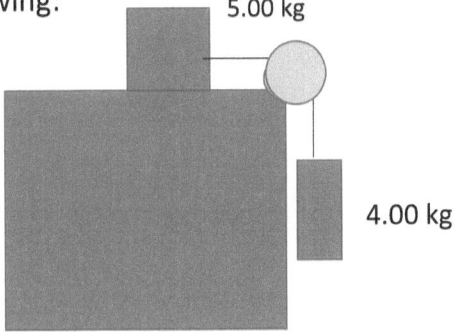

5.00 kg

4.00 kg

A...Net Force on the System:

B...Acceleration of the System:

C...Impulse at the end of 0.5 s:

D...Final Velocity of the System after 0.5s:

5...For the system below the hanging mass is 9.00 kg and the mass over the frictionless table is 8.00 kg. Find the following:

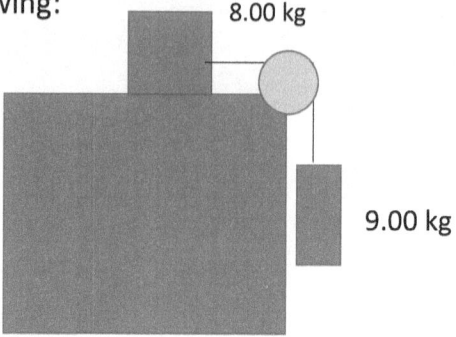

8.00 kg

9.00 kg

A...Net Force on the System:

B...Acceleration of the System:

C...Impulse at the end of 0.7 s:

D...Final Velocity of the System after 0.7s:

6...A person applies a variable force as shown in the graph below. Answer the following:

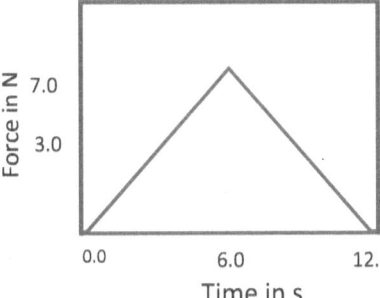

A...What is the Impulse on the box?

B...If the box is 9.8 kg what is its gain in Kinetic Energy if there is no Friction?

C...What is the Final Velocity?

D...What is the Average Acceleration?

E...What is the Power delivered to the box?

7...A person applies a variable force as shown in the graph below. Answer the following:

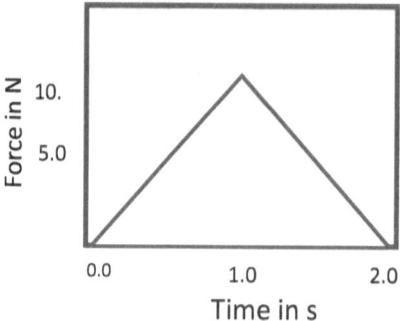

A...What is the Impulse on the box?

B...If the box is 2.8 kg what is its gain in Kinetic Energy if there is no Friction?

C...What is the Final Velocity?

D...What is the Average Acceleration?

E...What is the Power delivered to the box?

8...A person applies a variable force as shown in the graph below. Answer the following:

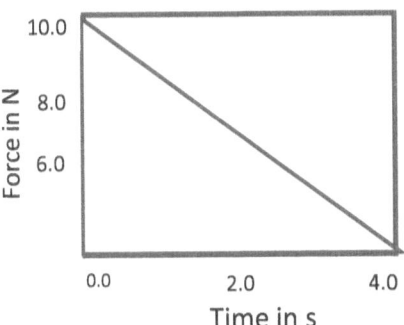

A...What is the Impulse on the box?

B...If the box is 9.2 kg what is its gain in Kinetic Energy if there is no Friction?

C...What is the Final Velocity?

D...What is the Average Acceleration?

E...What is the Power delivered to the box?

9...A person applies a variable force as shown in the graph below. Answer the following:

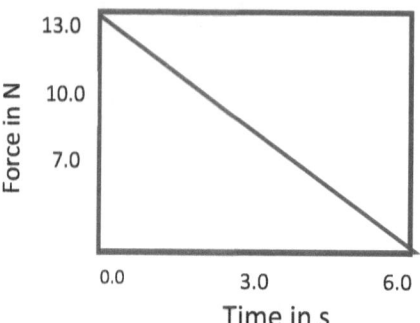

A...What is the Impulse on the box?

B...If the box is 2.0 kg what is its gain in Kinetic Energy if there is no Friction?

C...What is the Final Velocity?

D...What is the Average Acceleration?

E...What is the Power delivered to the box?

10...A person applies a variable force as shown in the graph below. Answer the following:

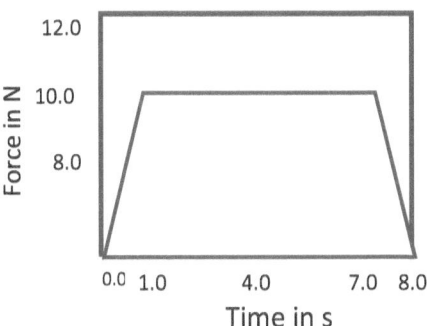

A...What is the Impulse on the box?

B...If the box is 0.9 kg what is its gain in Kinetic Energy if there is no Friction?

C...What is the Final Velocity?

D...What is the Average Acceleration?

E...What is the Power delivered to the box?

11...A person applies a variable force as shown in the graph below. Answer the following:

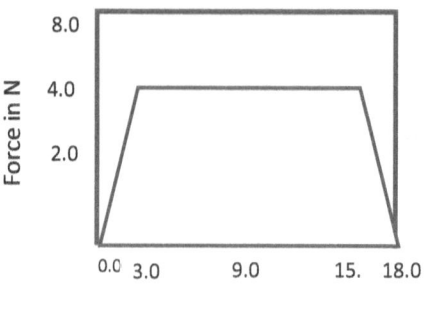

Time in s

A...What is the Impulse on the box?

B...If the box is 8.0 kg what is its gain in Kinetic Energy if there is no Friction?

C...What is the Final Velocity?

D...What is the Average Acceleration?

E...What is the Power delivered to the box?

Name_____Period_____

Momentum and Impulse Test

1...A Particle of mass 14.0 kg moves with Velocity 7.00 m/s to the right. It then enters a Field in which it experiences a Force of 7.00 N to the left for 7.00 s.

The name of the Particle is LaydeZerink!

A...What is the Initial Momentum of the Particle?

B...What is the Impulse of the Particle?

C...What is the change in Momentum of the Particle?

D...What is the Final Velocity of the Particle?

E...What is the gain in Kinetic Energy of the Particle?

F...What was the Acceleration of the Particle in that time?

G...What was the Work done on the Particle?

H....What is the Power that the Force delivered to LaydeZerink in that time?

2...For the system below the hanging mass is 3.00 kg and the mass over the frictionless table is 3.00 kg. Find the following:

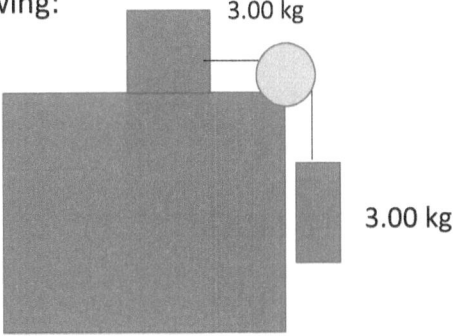

A...Net Force on the System:

B...Acceleration of the System:

C...Impulse at the end of 0.5 s:

D...Final Velocity of the System after 0.5s:

3...A person applies a variable force as shown in the graph below. Answer the following:

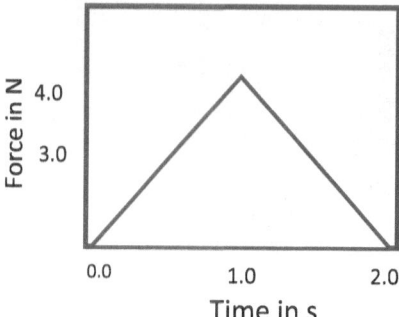

A...What is the Impulse on the box?

B...If the box is 3.8 kg what is its gain in Kinetic Energy if there is no Friction?

C...What is the Final Velocity?

D...What is the Average Acceleration?

E...What is the Power delivered to the box?

4...A person applies a variable force as shown in the graph below. Answer the following:

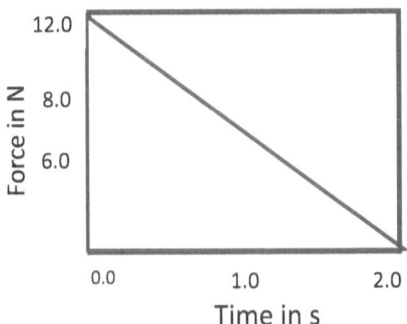

A...What is the Impulse on the box?

B...If the box is 1.2 kg what is its gain in Kinetic Energy if there is no Friction?

C...What is the Final Velocity?

D...What is the Average Acceleration?

E...What is the Power delivered to the box?

5...A person applies a variable force as shown in the graph below. Answer the following:

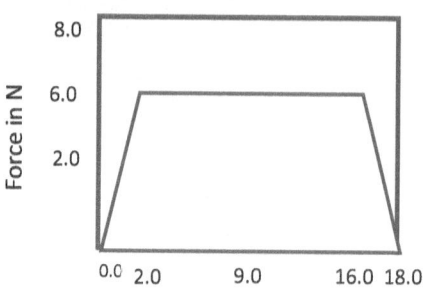

Time in s

A...What is the Impulse on the box?

B...If the box is 4.0 kg what is its gain in Kinetic Energy if there is no Friction?

C...What is the Final Velocity?

D...What is the Average Acceleration?

E...What is the Power delivered to the box?

6...A 10.0 kg ball moves with a Velocity of 7.00 m/s to the right.

A...What is its Momentum?

B...The ball collides with another ball of mass 10.0 kg and then stops. What is the Velocity of the second ball?

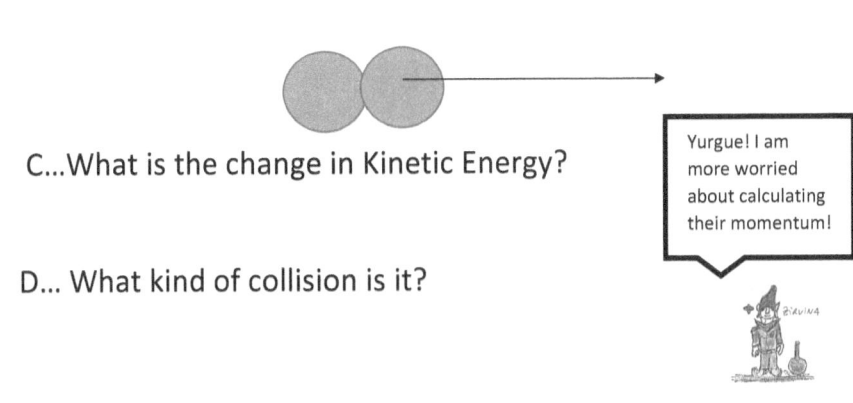

C...What is the change in Kinetic Energy?

D... What kind of collision is it?

7...A 10.0 kg ball moves with a Velocity of 9.00 m/s to the right.

Zirvina! Particle collisions are also conceptually amazing!

A...What is its Momentum?

B...The ball collides with a 10.0 kg ball and they both stick together. What is the Velocity of both moving together?

C...What is the change in Kinetic Energy?

D... What kind of collision is it?

8...A 1.0 kg soccer ball is moving with a Velocity of 30 m/s to the right towards a wall. It hits the wall and experiences a Force depicted in the graph below:

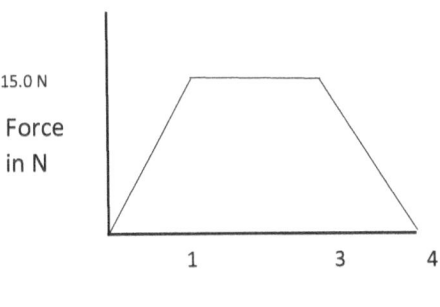

Time in s

A...What is the Impulse on the ball?

Please place a negative sign in front of the area under the curve: _____

B...What is the change in Momentum of the ball?

C...What is the Final Velocity of the ball? _____

Use: - (Area under curve) = m (Vf – 30)

9...A 1.00 kg soccer ball is dropped from a building that is 100 m:

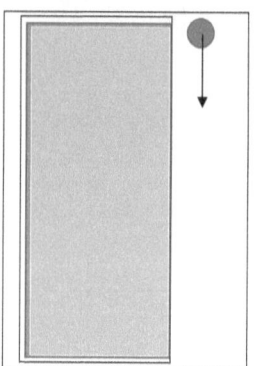

A... What is the Final Momentum of the ball when it hits the ground?

B... The ball hits the ground and comes to rest at 0.5 ms. What is the Force that the ground exerted on the ball?

The strangest thing about the universe is that it is comprehensible.

Name_____ Period_____

Collisions

1. A block of mass 1.00 kg moves with a Velocity of 2.00 m/s. It then collides with a block of mass 3.00 kg. After collision, the 1.00 kg block bounces back -1.00 m/s.

   A. With what Velocity does the 3.00 kg block moves with both magnitude and direction?

   B. What is the Initial Kinetic Energy before collision?

   C. What is the Final Kinetic Energy?

   D. When the collision is Elastic, Kinetic Energy is conserved. When the collision is Inelastic, Kinetic Energy is not conserved. What kind of collision is this?

2. A block of mass 1.00 kg moves with a Velocity of 2.00 m/s. It then collides with a block of 3.00 kg. After the collision the 3.00 kg block moves with a Velocity of 0.833 m/s. What is the Velocity of the 1.00 kg block?

B. What is the Initial Kinetic Energy of the system before collision?

C. What is the Final Kinetic Energy of the system?

D. What kind of collision is it?

E. Was Kinetic Energy lost or gained?

3. A 1.00 kg block moves with Velocity of 2.00 m/s. It collides with a 2.00 kg that is moving at -2.00 m/s in a head on collision.

A. After collision, the 1.00 kg block bounced back at -3.00 m/s. With what Velocity did the 2.00 kg block forward?

B. What is the Initial Kinetic Energy of the system before collision?

C. What is the Final Kinetic Energy or the system after collision?

D. What kind of collision is it?

Name_____ Period____

Momentum and Impulse Review

1…A 5.60 kg soccer ball is moving at 10.0 m/s. A player strikes that ball that leaves his foot at 20.0 m/s in the opposite direction. The impact lasted 0.200s.

A…What is the average force that the ball exerted on the players foot?

B…What is the Impulse on the ball?

C…What was the average acceleration of the ball during impact?

D…The ball is then grabbed by a 75.0 kg goalkeeper who is stepping on a frictionless surface. What is the final velocity of the ball and goalkeeper?

E…What is the loss in kinetic energy?

F…The goalkeeper and the ball then hit a post and they both come to stop at 0.300 s. What is the impulse on both?

G…What is the force that the post exerted on the goalkeeper holding the ball?

H...A 56.0 cart is passing by the goalkeeper at 5.00 m/s. The goalkeeper drops the ball there. What is the final velocity of the cart holding the ball inside?

I...What is the loss in kinetic energy?

2...A 78.0 kg astronaut initially at rest throws a 6.00 kg tool in one direction at 20.0 m/s propelling him backwards.

A...What is the velocity of the astronaut afterwards?

B...What is the impulse on the astronaut?

C...What is the gain in kinetic energy of the system?

D...The astronaut then hits the spaceship and comes to a stop in 0.500s. What is his impulse?

E...What is the average force that acted on the astronaut during collision?

F…What is the average acceleration that acted on the astronaut during collision?

G…Explain where the loss of kinetic energy go:

3…The area under a force and displacement curve is the_____.

4…The area under a force and time interval curve is the_____.

5…Find the impulse on an object with the following graph of force versus time interval:

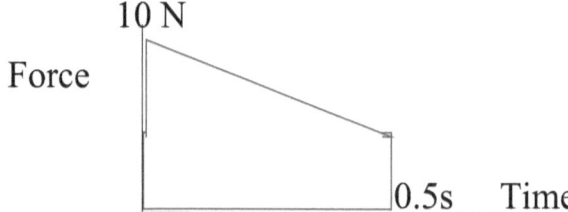

Name_____Period____

Thermal Physics

1...34.0 kg of ice is heated from $-7.0°C$ to $0°C$:

A...How much heat must be added to the ice?

B...If the heater generates 600 W of Power, how many minutes will it take?

2...The Ice now melts.

A... How much heat must be added to the ice?

B... If the heater generates 10,000 W of Power, how many minutes will it take?

3...The liquid water is heated to $50°C$.

A... How much heat must be added to the ice?

B... If the heater generates 5,000 W of Power, how many minutes will it take?

4...The liquid water's temperature drops to $20°C$.

A...How much heat was lost?

B...How many minutes did it take to cool if the rate of cooling was 500 W?

5...How much heat will be needed to raise the liquid water's temperature to $100°C$?

6..How much heat will it take to vaporize this liquid water?

7...What must be the power output to vaporize the liquid water in 15 minutes?

8...How much heat must be taken out of the water to cool it from vapor to 22°C?

9...Fill the Graph below with the labels:

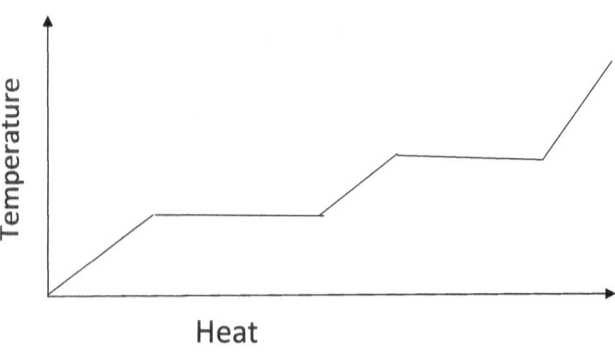

Heat

Label the graph with Gas, Liquid, and Solid Phase and Phase Changes

10...67.0 kg of water at $3.0°C$ is dropped in 85.0 kg of liquid water at $50°C$:

A...What will be the Final Temperature of the mixture?

11...An Ideal Gas at 297K, 8.0 m^3 Volume, and 5.00×10^5 Pa of Pressure.

A...The Volume is reduced to a half. What is the new Pressure if Temperature is kept the same?

B...How many moles of gas are there?

C...How many atoms of gas are there?

D...By raising the Temperature to 303K, and keeping the Pressure the same, what is the new Volume?

E...What is the Internal Energy of the gas?

F...What is the Root Mean Square Velocity of the atoms in the gas? (Molar Mass is 1.00 g)

12...17.0 kg of water at $9.0°C$ is dropped in 25.0 kg of liquid water at $30°C$:

A...What will be the Final Temperature of the mixture?

13...10.0 kg of water at $4.0°C$ is dropped in 25.0 kg of liquid water at $10°C$:

A...What will be the Final Temperature of the mixture?

14...State what happens to Kinetic Energy and Potential Energy when the following happens:

Change	Kinetic Energy	Potential Energy
Temperature Raised		
Temperature Dropped		
Phase Change Heat In		
Phase Change Heat Out		
No change		

15...What is Internal Energy?

16...What defines an Ideal Gas?

17...The bottom graph shows the Speed of Molecules in an Ideal Gas at different Temperatures:

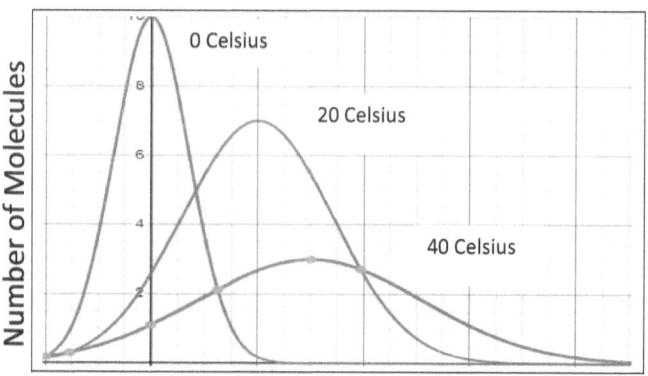

Speed of Molecules

A...State 4 things observed in this graph:

1.

2.

3.

4.

18...Label in the graph below the regions for Melting, Freezing, Boiling, and Condensation:

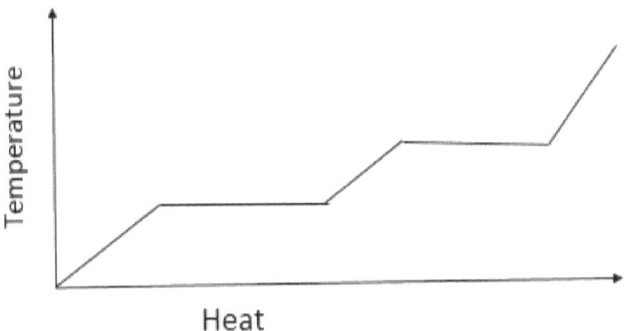

19...In the image below label the names of phase changes:

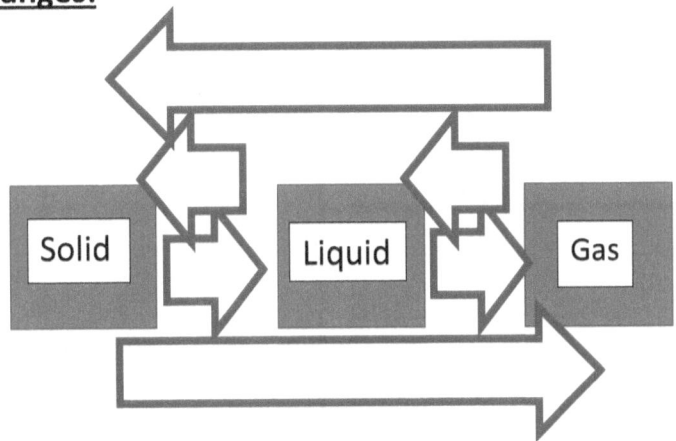

Gas Laws Graphs:

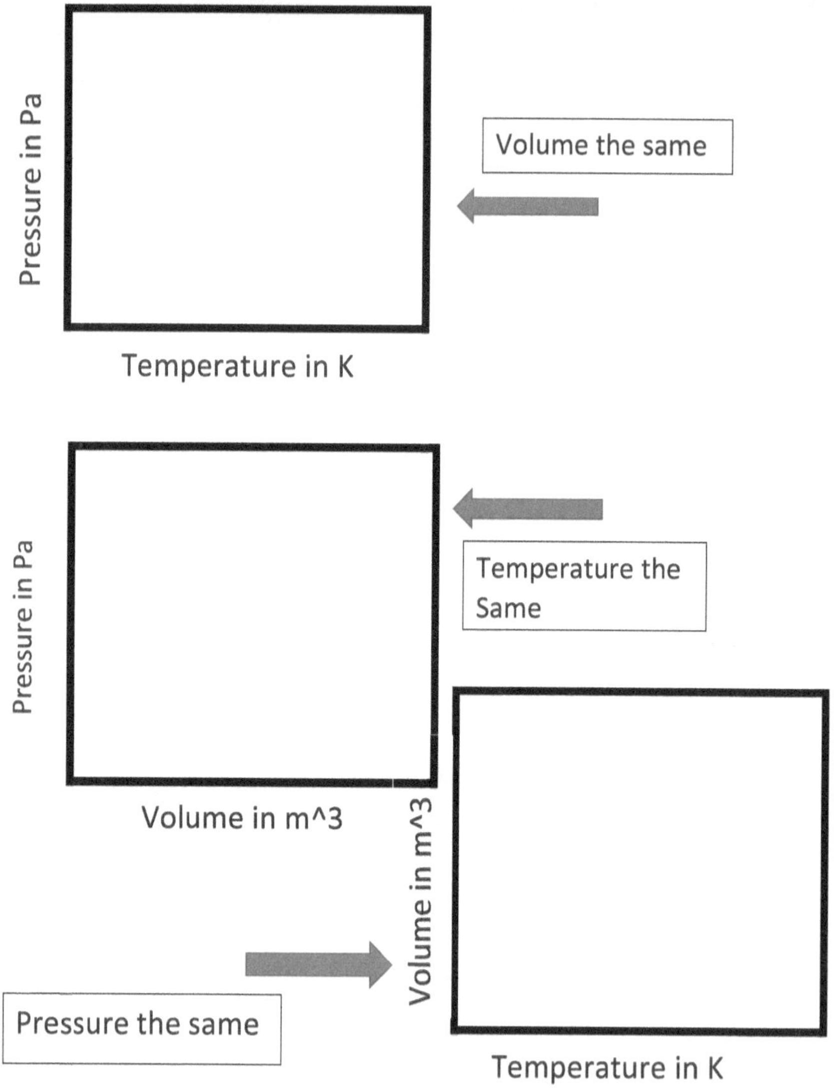

Pressure in Pa | Temperature in K

← Volume the same

Pressure in Pa | Volume in m^3

← Temperature the Same

Volume in m^3 | Temperature in K

← Pressure the same

Name_____ Period____

Thermal Physics 2

1...84.0 kg of ice is heated from $-13.0°C$ to $0°C$:

A...How much heat must be added to the ice?

B...If the heater generates 900 W of Power, how many minutes will it take?

2...The Ice now melts.

A... How much heat must be added to the ice?

B... If the heater generates 40,000 W of Power, how many minutes will it take?

3...The liquid water is heated to $70°C$.

A... How much heat must be added to the ice?

B... If the heater generates 8,000 W of Power, how many minutes will it take?

4...The liquid water's temperature drops to $10°C$.

A...How much heat was lost?

B...How many minutes did it take to cool if the rate of cooling was 900 W?

5…How much heat will be needed to raise the liquid water's temperature to $100°C$?

6..How much heat will it take to vaporize this liquid water?

7…What must be the power output to vaporize the liquid water in 35 minutes?

8...How much heat must be taken out of the water to cool it from vapor to $3°C$?

9...Fill the Graph below with the labels:

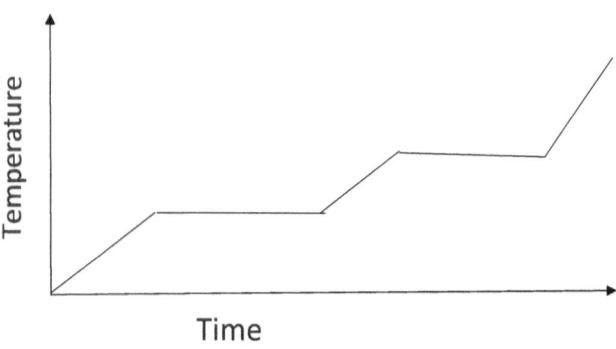

Label the graph with Gas, Liquid, and Solid Phase and Phase Changes

10...17.0 kg of water at $7.0°C$ is dropped in 75.0 kg of liquid water at $70°C$:

A...What will be the Final Temperature of the mixture?

11...An Ideal Gas at 300K, 9.0 m^3 Volume, and 8.00×10^5 Pa of Pressure.

A...The Volume is reduced to a half. What is the new Pressure if Temperature is kept the same?

B...How many moles of gas are there?

C...How many atoms of gas are there?

D...By raising the Temperature to 400K, and keeping the Pressure the same, what is the new Volume?

E...What is the Internal Energy of the gas?

F...What is the Root Mean Square Velocity of the atoms in the gas? (Molar Mass 7g)

12...27.0 kg of water at $29.0°C$ is dropped in 55.0 kg of liquid water at $80°C$:

A...What will be the Final Temperature of the mixture?

13...10.0 kg of water at $5.0°C$ is dropped in 45.0 kg of liquid water at $30°C$:

A...What will be the Final Temperature of the mixture?

Name_____Period_____

Thermal Review 3

A...A 8.00 kg of ice at -5.0 °C is dropped in 74.0 kg liquid water at 80°C. Find the following:

    Final temperature of the mixture:

    Number of moles of water:

    Number of water molecules:

B...A 2.00 kg of ice at -7.0 °C is dropped in 84.0 kg liquid water at 60°C. Find the following:

Final temperature of the mixture:

Number of moles of water:

Number of water molecules:

C...A 2.00 kg of water at 6.0 °C is dropped in 44.0 kg liquid water at 20°C. Find the following:

Final temperature of the mixture:

Number of moles of water:

Number of water molecules:

Name_____ Period_____

Notes on Waves 1

According to Einstein's most famous equation, Energy equals to Mass times the Speed of Light squared:

$$E = mc^2$$

This equation states that Energy can be converted to Mass and Mass can be converted to Energy. Everything in the entire universe is Energy.

In space among its multiples Quantum Fluctuations, Particles and Anti-Particles are created from Energy, exist for a brief amount of time, and annihilate becoming Energy again.

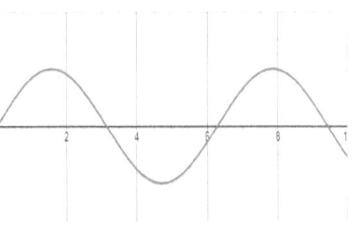

These quantum Fluctuations exist throughout space in what is known as Hilbert Space. The sum of all of these Waves or Oscillations gives form to the universe. All matter in the cosmos is composed of the sum of all the Waves present in a region of space.

When studying Waves or Oscillations, Physicists discovered Simple Harmonic Motion which is a type of repetitive or periodical motion that can be described with Sinusoidal Waves such as Sine and Cosine.

Below is a Sine Wave.

Label the Amplitude, and Wavelength:

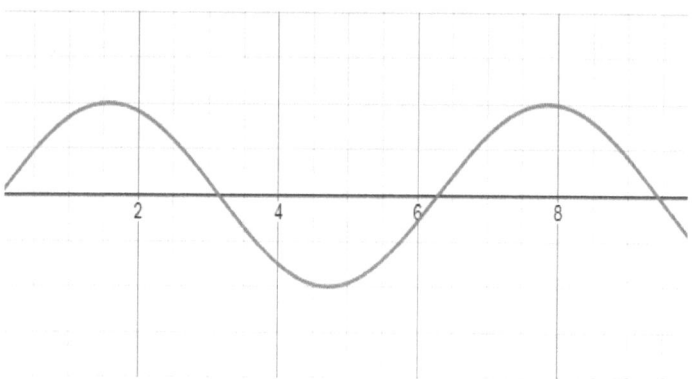

Any oscillation or repetitive motion can be described with Waves. Cycles can be measured with Periods, Frequencies, and Angular Frequencies.

A Period is how many seconds in a cycle and is measured in seconds.

A Frequency is how many cycles per second and is measured in Hertz.

Angular Frequency is Angles per seconds and is measured in Radians per second.

1...Calculate the Period and Frequency for the following:

Situation	Period (s)	Frequency (Hz)
The seconds hand of a clock		
The minutes hand of a clock		
The hours hand of a clock		

The rotation of Earth once around its own axis		
The rotation of Earth around the Sun		
The Rotation of the Moon around the Earth		

Two cases worth exploring when studying Waves is the Pendulum and the Spring Mass System.

Let us investigate first the Pendulum:

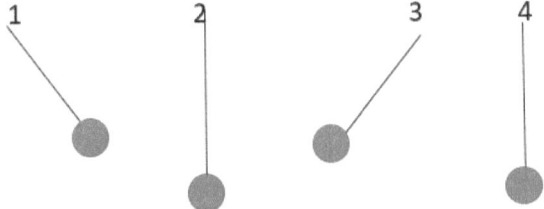

If the Length of the Pendulum is known its Period can be calculated as:

$$T = 2\pi \sqrt{\frac{L}{g}}$$

The Frequency of any oscillation is found by taking the inverse of the Period:

$$Frequency = \frac{1}{Period}$$

Situation Length of Pendulum	Period	Frequency
1.00 m		
0.09 m		
3.35 m		
4.32 m		
0.15 m		

Derive an Equation is terms of Length and Period of a Pendulum to solve for the Acceleration due to Gravity:

Do the same calculation for a Pendulum on the Moon with Gravity equal to 1/6 that of the Earth:

Situation Length of Pendulum	Period	Frequency
1.00 m		
0.09 m		
3.35 m		

4.32 m		
0.15 m		

The other case is that of a Spring Mass System.

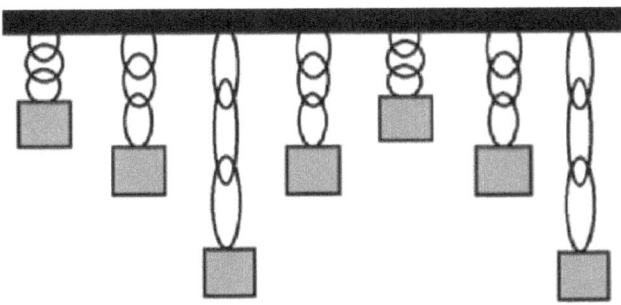

If the Mass of the Hanging Mass and the Spring Constant are known the Period can be found with the following equation:

The Frequency of any oscillation is found by taking the inverse of the Period:

$$T = 2\pi \sqrt{\frac{m}{k}}$$

$$\text{Frequency} = \frac{1}{Period}$$

Situation Mass of Hanging Mass	Period	Frequency
1.00 kg K =7N/m		
0.09 kg K = 6 N/m		
3.35 kg K=9N/m		
4.32kg K=0.1N/m		
0.15kg K=9.8N/m		

State where in the Oscillation of a Pendulum is the Kinetic and Potential Energies maximum or zero:

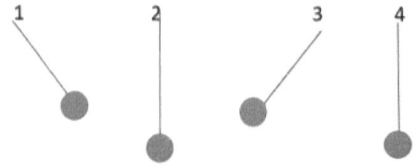

Max Potential Energy:

Max Kinetic Energy:

Zero Potential Energy:

Zero Kinetic Energy:

State where is in the Oscillation of a Spring is the Kinetic and Potential Energies maximum or zero:

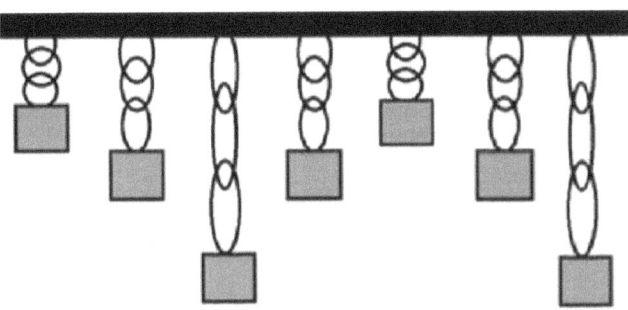

Max Potential Energy:
Max Kinetic Energy:
Zero Potential Energy:
Zero Kinetic Energy:

Oscillations Review:

1-What is the Period of Wave that makes 10 cycles per second?

2-What is the Frequency of the Wave in problem 1?

3-What is the Period of a Spring Mass System with k=10.0 N/m and mass = 56g?

4-What is the Angular Frequency of the system in problem 3?

5-What is the Frequency of the system in problem 3?

6-What is the Length of Pendulum with a Period of 0.32 seconds?

7-What is the Acceleration due to Gravity on a Planet if a 10 m long Pendulum has a Period of 1.0 second?

8- Indicate in the above figure the Amplitude, the Wavelength, and where the equilibrium position is.

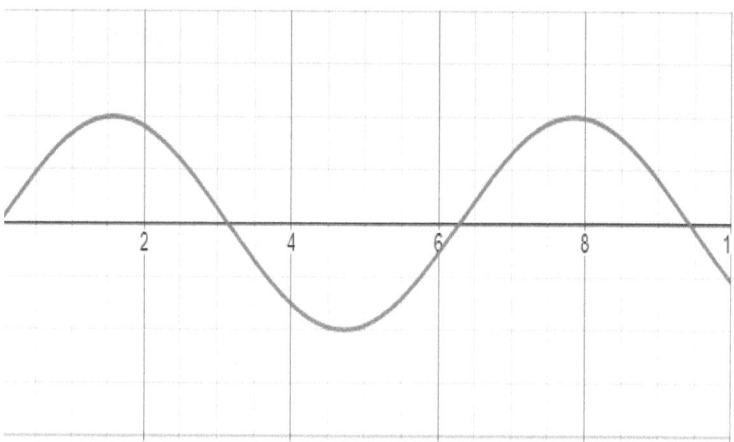

9- What is the Energy stored in a Spring with k = 100N/m compressed 26 cm?

10- What is the Restoring Force in the spring on problem 9?

11- What is Spring Constant?

12- How long should a Pendulum be to have the same Period as the Spring Mass System on problem 3?

13- What is Restoring Force?

14- At what location in a Simple Harmonic Motion is the Velocity the greatest?

15- Where is the Velocity zero?

16- What is the Velocity of oscillation at the farthest position from Equilibrium?

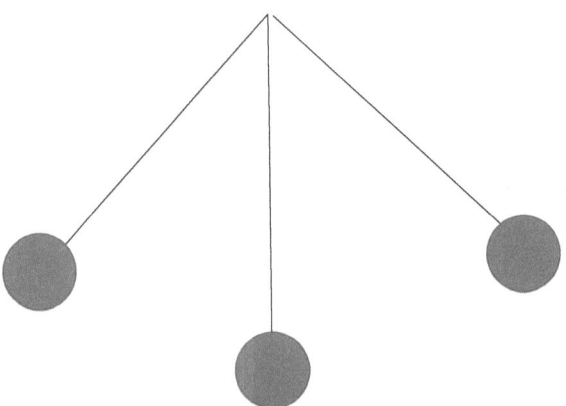

17- In the figure above indicate the Equilibrium Position, and the Maximum Displacement.

Name_____Period_____

Oscillation Review 2

1...A Spring with Spring Constant K= 4.0 N/m has a length of 4.0 m. A 3.00 kg box is about to be hanged of the spring:

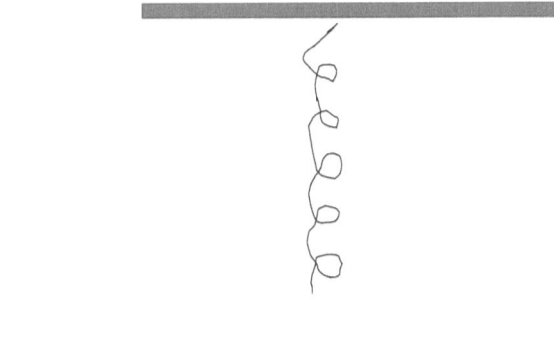

A....When the block hangs on the Spring, what is the amount of stretch?

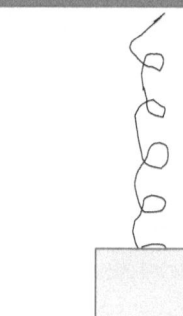

B...What is the Period if it begins to oscillate?

C...What is the Frequency?

D...What is the Angular Frequency?

E...What is the Total Energy of Oscillation?

F...What is the Maximum Speed of Oscillation?

2...A Spring with Spring Constant K= 3.0 N/m has a length of 2.0 m. A 2.00 kg box is about to be hanged of the spring:

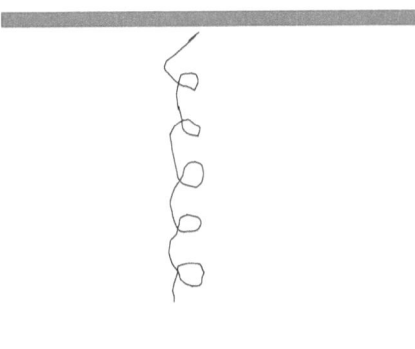

A....When the block hangs on the Spring, what is the amount of stretch?

B...What is the Period if it begins to oscillate?

C...What is the Frequency?

D...What is the Angular Frequency?

E...What is the Total Energy of Oscillation?

F...What is the Maximum Speed of Oscillation?

3...A Spring with Spring Constant K= 7.0 N/m has a length of 6.0 m. A 1.00 kg box is about to be hanged of the spring:

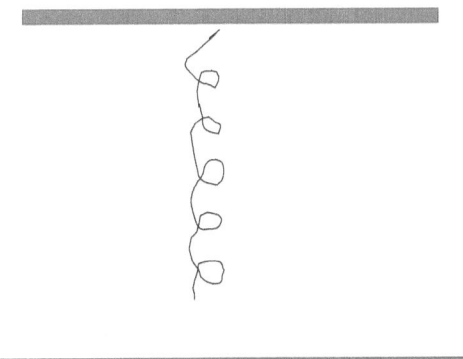

A....When the block hangs on the Spring, what is the amount of stretch?

180

B…What is the Period if it begins to oscillate?

C…What is the Frequency?

D…What is the Angular Frequency?

E…What is the Total Energy of Oscillation?

F…What is the Maximum Speed of Oscillation?

For the Pendulum answer the following questions: | Mass of bob = 7.00 kg

1 2 3 4

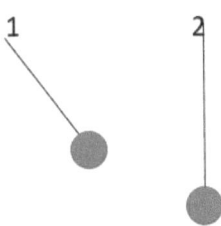

The length of the Pendulum is 1.00 m

4...Where in the cycle is:

Kinetic Energy Max:_____

Potential Energy Max:_____

Kinetic Energy zero:_____

Potential Energy zero:_____

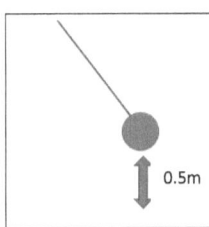

0.5m

5...If the Pendulum was dropped when the Bob was at a height of 0.5 m from its Equilibrium, how fast will it be moving when Kinetic Energy is Max?

What is the Total Energy of this oscillating Pendulum?_____

6...How fast will it be moving at Max Potential Energy? _____

7...If after each cycle the Pendulum loses 25% of its Energy, how fast will it be moving at Equilibrium Position after 1 cycle?

8.....What is the Period of the Pendulum?

9...What is the Frequency of the Pendulum?

10...What is the Angular Frequency of the Pendulum?

11...Label the parts of the cycle in the Cosine graph below:

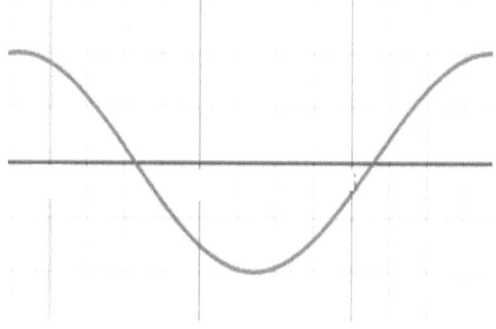

Left position is Negative Amplitude and right position is Positive Amplitude

For the Pendulum answer the following questions: | Mass of bob = 2.00 kg

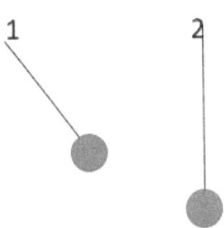

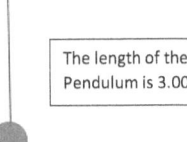

The length of the Pendulum is 3.00 m

12...Where in the cycle is:

Kinetic Energy Max:_____

Potential Energy Max:_____

Kinetic Energy zero:_____

Potential Energy zero:_____

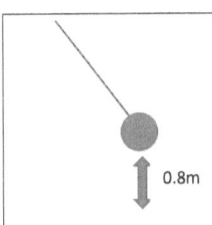

13...If the Pendulum was dropped when the Bob was at a height of 0.8 m from its Equilibrium, how fast will it be moving when Kinetic Energy is Max?

What is the Total Energy of this oscillating Pendulum?_____

14..How fast will it be moving at Max Potential Energy? _____

15.If after each cycle the Pendulum loses 25% of its Energy, how fast will it be moving at Equilibrium Position after 1 cycle?

16...What is the Period of the Pendulum?

17...What is the Frequency of the Pendulum?

18...What is the Angular Frequency of the Pendulum?

19...Label the parts of the cycle in the Cosine graph below:

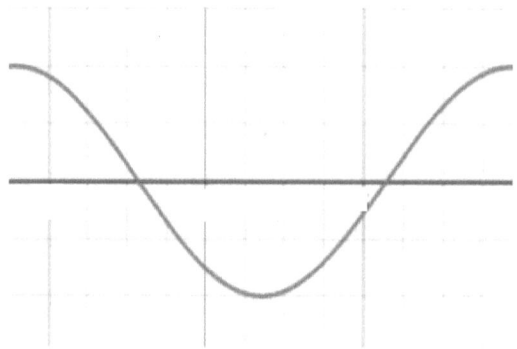

Left position is Negative Amplitude and right position is Positive Amplitude

Name_____ Period____

Oscillation Review 3

1…A Spring with Spring Constant K= 8.0 N/m has a length of 8.0 m. A 7.00 kg box is about to be hanged of the spring:

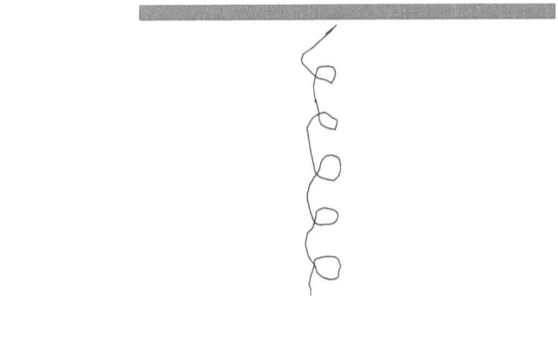

A….When the block hangs on the Spring, what is the amount of stretch?

B…What is the new length of the spring?

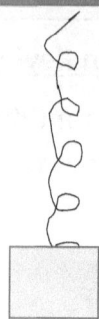

C…What is the Period if it begins to oscillate?

D…What is the Frequency?

E…What is the Angular Frequency?

F…What is the Total Energy of Oscillation?

G…What is the Maximum Speed of Oscillation?

2...A Spring with Spring Constant K= 1.0 N/m has a length of 1.0 m. A 4.00 kg box is about to be hanged of the spring:

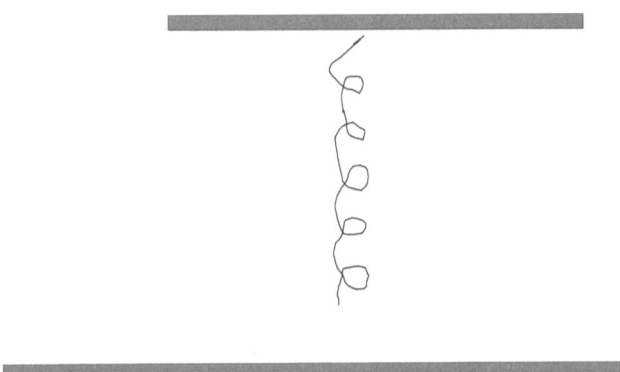

A....When the block hangs on the Spring, what is the amount of stretch?

B...What is the new length of the spring?

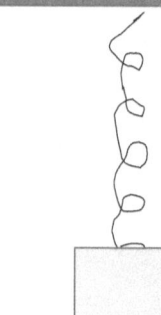

C...What is the Period if it begins to oscillate?

D...What is the Frequency?

E...What is the Angular Frequency?

F...What is the Total Energy of Oscillation?

G...What is the Maximum Speed of Oscillation?

3...A Spring with Spring Constant K= 5.0 N/m has a length of 3.0 m. A 2.00 kg box is about to be hanged of the spring:

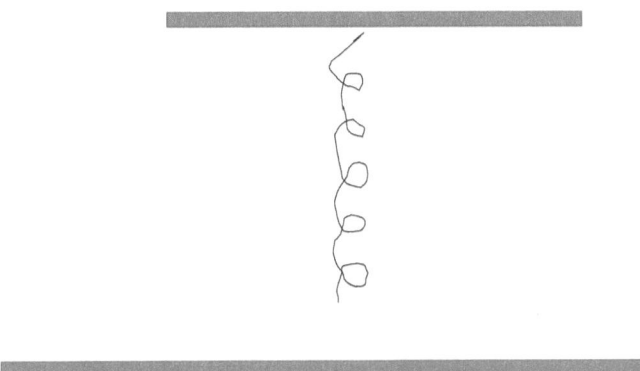

A....When the block hangs on the Spring, what is the amount of stretch?

B...What is the new length of the spring?

C…What is the Period if it begins to oscillate?

D…What is the Frequency?

E…What is the Angular Frequency?

F…What is the Total Energy of Oscillation?

G…What is the Maximum Speed of Oscillation?

For the Pendulum answer the following questions: | Mass of bob = 3.00 kg |

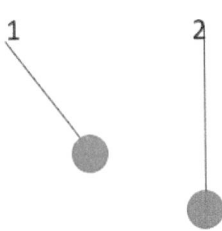

The length of the Pendulum is 4.00 m

4...Where in the cycle is:

Kinetic Energy Max:_____

Potential Energy Max:_____

Kinetic Energy zero:_____

Potential Energy zero:_____

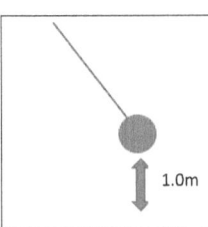

1.0m

5...If the Pendulum was dropped when the Bob was at a height of 1.0 m from its Equilibrium, how fast will it be moving when Kinetic Energy is Max?

What is the Total Energy of this oscillating Pendulum?_____

6...How fast will it be moving at Max Potential Energy? _____

7...If after each cycle the Pendulum loses 35% of its Energy, how fast will it be moving at Equilibrium Position after 1 cycle?

8.....What is the Period of the Pendulum?

9...What is the Frequency of the Pendulum?

10...What is the Angular Frequency of the Pendulum?

11...Label the parts of the cycle in the Cosine graph below:

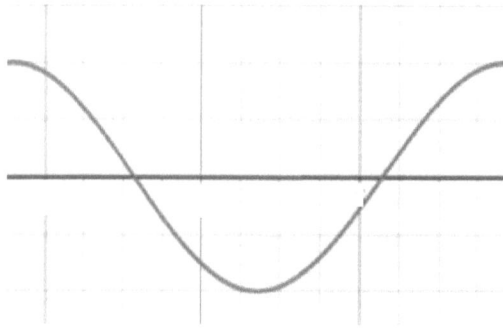

Left position is Negative Amplitude and right position is Positive Amplitude

For the Pendulum answer the following questions: Mass of bob = 7.00 kg

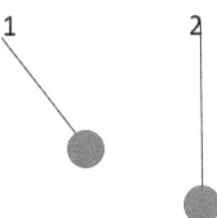
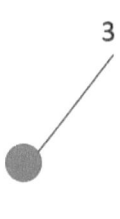
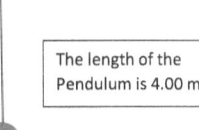

The length of the Pendulum is 4.00 m

12…Where in the cycle is:

Kinetic Energy Max:_____

Potential Energy Max:_____

Kinetic Energy zero:_____

Potential Energy zero:_____

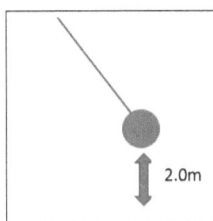

2.0m

13…If the Pendulum was dropped when the Bob was at a height of 2.0 m from its Equilibrium, how fast will it be moving when Kinetic Energy is Max?

What is the Total Energy of this oscillating Pendulum?_____

14..How fast will it be moving at Max Potential Energy? _____

15.If after each cycle the Pendulum loses 45% of its Energy, how fast will it be moving at Equilibrium Position after 1 cycle?

16…What is the Period of the Pendulum?

17...What is the Frequency of the Pendulum?

18...What is the Angular Frequency of the Pendulum?

19...Label the parts of the cycle in the Cosine graph below:

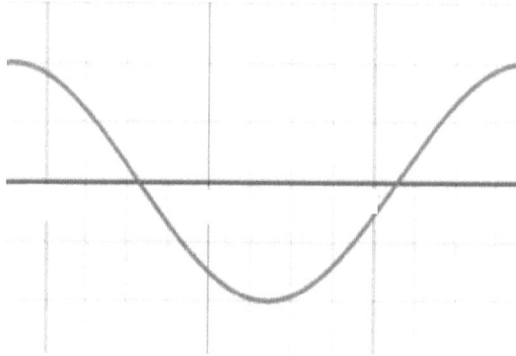

Left position is Negative Amplitude and right position is Positive Amplitude

Oscillations Review:

20-What is the Period of Wave that makes 30 cycles per second?

21-What is the Frequency of the Wave in problem 20?

22-What is the Period of a Spring Mass System with k=5.0 N/m and mass = 16g?

23-What is the Angular Frequency of the system in problem 22?

24-What is the Frequency of the system in problem 22?

25-What is the Length of Pendulum with a Period of 0.50 seconds?

26-What is the Acceleration due to Gravity on a Planet if a 10 m long Pendulum has a Period of 5.00 seconds?

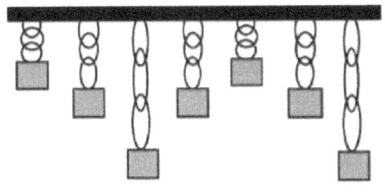

For the Spring Mass System of the left answer the following questions:

If the spring has K= 9.00 N/m, the hanging mass is 1.80 kg, and is Displaced 0.5 m:

1...Total energy of the oscillation:

2...State the locations of:

Max Potential Energy:_____

3...Where is the Max speed is reached?

Max Kinetic Energy:_____

Zero Potential Energy:_____

4...What is the Max speed?

Zero Kinetic Energy:_____

8...What is the Period of oscillation?

5...Where is the Speed zero?

6...If the Spring Mass System loses 20% of its energy is each oscillation, what is the Max Speed after 3 cycles?

9...What is the Frequency of oscillation?

7...How much will it compress and stretch on the third cycle?

10...What is the Angular Frequency of oscillation?

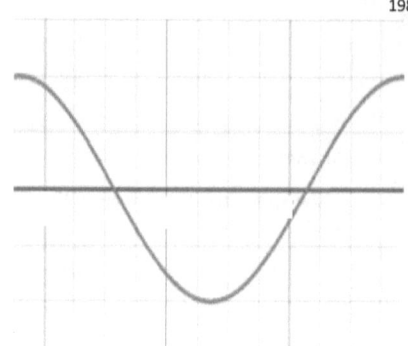

11...Label the parts of the cycle in the Cosine graph on the left:

Up Position is Positive Amplitude and Down Position is Negative Amplitude

Oscillations questions:

1-What is the Period of Wave that makes 900 cycles per hour?

2-What is the Frequency of the Wave in problem 1?

3-What is the Period of a Spring Mass System with k=90.0 N/m and mass = 1.00kg?

4-What is the Angular Frequency of the system in problem 3?

5-What is the Frequency of the system in problem 3?

6-What is the Length of Pendulum with a Period of 0.25 seconds?

7-What is the Acceleration due to Gravity on a Planet if a 1.50 m long Pendulum has a Period of 2.00 seconds?

8-How many times Earth's Gravity is this Planet?

9-What must be its Radius if its Mass is the same of the Earth?

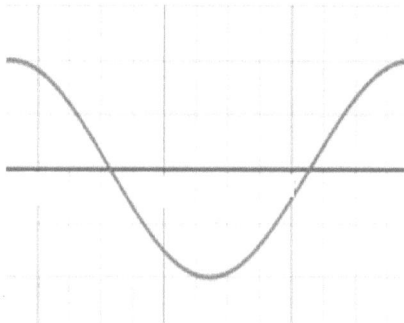

10- Indicate in the above figure the Amplitude, the Wavelength, where the equilibrium position is, the crests, and troughs.

Name_____Period____

Oscillations Test

1...A Spring with Spring Constant K= 3.0 N/m has a length of 3.0 m. A 3.00 kg box is about to be hanged of the spring:

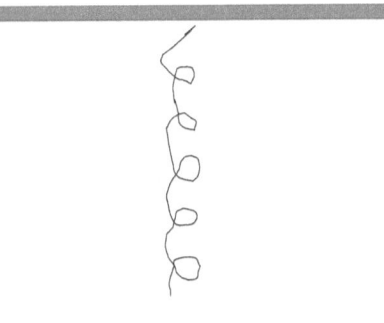

A....When the block hangs on the Spring, what is the amount of stretch?

B...What is the new length of the spring?

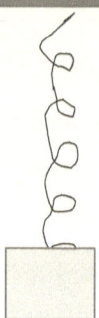

C...What is the Period if it begins to oscillate?

D...What is the Frequency?

E...What is the Angular Frequency?

F...What is the Total Energy of Oscillation?

G...What is the Maximum Speed of Oscillation?

2...A Spring with Spring Constant K= 2.0 N/m has a length of 2.2 m. A 2.22 kg box is about to be hanged of the spring:

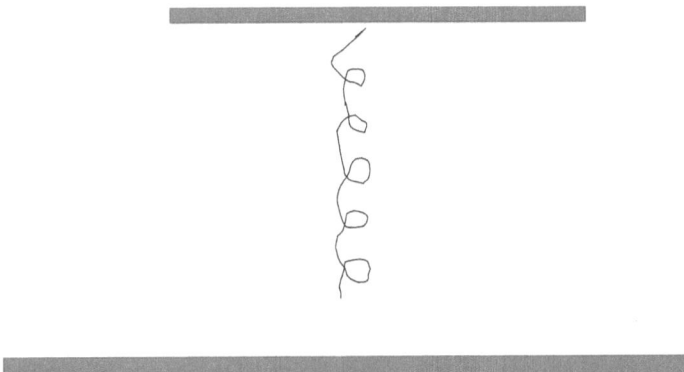

A....When the block hangs on the Spring, what is the amount of stretch?

B...What is the new length of the spring?

C...What is the Period if it begins to oscillate?

D...What is the Frequency?

E...What is the Angular Frequency?

F...What is the Total Energy of Oscillation?

G...What is the Maximum Speed of Oscillation?

3..For the Pendulum answer the following questions Mass of bob = 2.50 kg

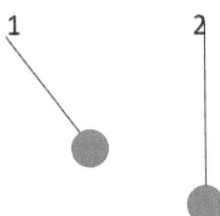

The length of the Pendulum is 5.80 m

4...Where in the cycle is:

Kinetic Energy Max:_____

Potential Energy Max:_____

Kinetic Energy zero:_____

Potential Energy zero:_____

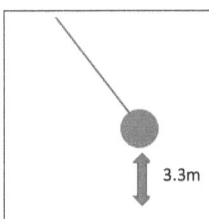

3.3m

5...If the Pendulum was dropped when the Bob was at a height of 3.3 m from its Equilibrium, how fast will it be moving when Kinetic Energy is Max?

What is the Total Energy of this oscillating Pendulum?_____

6...How fast will it be moving at Max Potential Energy? _____

7...If after each cycle the Pendulum loses 15% of its Energy, how fast will it be moving at Equilibrium Position after 1 cycle?

8.....What is the Period of the Pendulum?

9...What is the Frequency of the Pendulum?

10...What is the Angular Frequency of the Pendulum?

11...Label the parts of the cycle in the Cosine graph below:

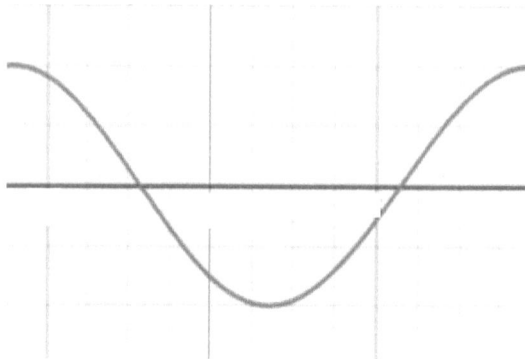

Left position is Negative Amplitude and right position is Positive Amplitude

Oscillations Review:

12- What is the Period of Wave that makes 20 cycles per second?

13- What is the Frequency of the Wave in problem 12?

14- What is the Period of a Spring Mass System with k=2.0 N/m and mass = 26g?

15- What is the Angular Frequency of the system in problem 14?

16-What is the Frequency of the system in problem 14?

17-What is the Length of Pendulum with a Period of 0.8 seconds?

18-What is the Acceleration due to Gravity on a Planet if a 8.0 m long Pendulum has a Period of 4.50 seconds?

Name_____Period____

Waves IB Physics

1. What is the Frequency of a Wave with a Period of 0.50 s?

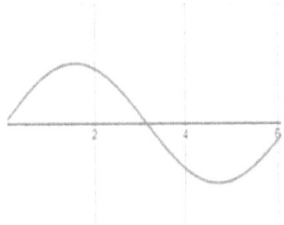

2. What is the Speed of the Wave if its Wavelength is 10.0 m?

3. Graph the Wave below: The Amplitude is 10.0 m:

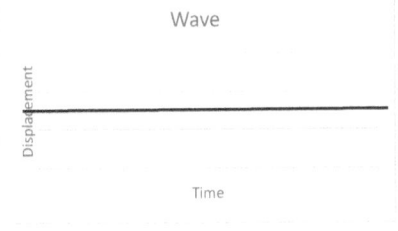

Wave / Displacement / Time

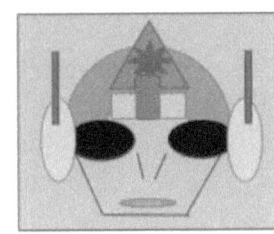

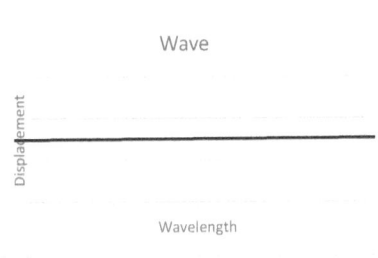

Wave / Displacement / Wavelength

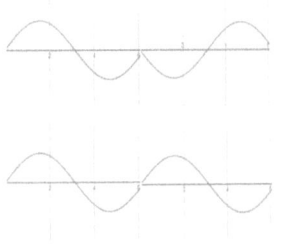

4. In the Oscillating Pendulum below answer the following:

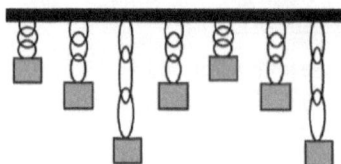

A. Indicate the locations of Maximum Potential Energy:

B. Indicate the Locations of Maximum Kinetic Energy:

C. The Spring Constant is 70.0 N/m. The Mass hanging on it is 78.0 kg. How much does the spring stretch on its own?

D. What is the Frequency the Oscillation after you pulled it 2.00 m:

$$f = \frac{1}{2\pi}\sqrt{\frac{k}{m}}$$

E. What is the Angular Frequency of the Oscillation?

F. Write the Equation for this wave:

$$\psi = A \cos(2\pi f t + \theta)$$

A = amplitude
f = frequency
θ = initial displacement

G. Graph this Wave and state the equation:

Wave	Equation:
Displacement vs Time	

Wave	Equation:
Velocity vs Time	

Wave	Equation:
Acceleration vs Time	

Waves are everywhere and the Universe is filled with them composing all the matter and energy of the cosmos.

H. Convert the Following:

Sin(θ)   into   Cos(θ)

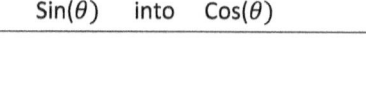

Sin(θ)   into   -Cos(θ)

Cos(θ)   into   Sin(θ)

Cos(θ)   into   -Sin(θ)

I. Describe the points in the circle that are represented in a Sine Wave:

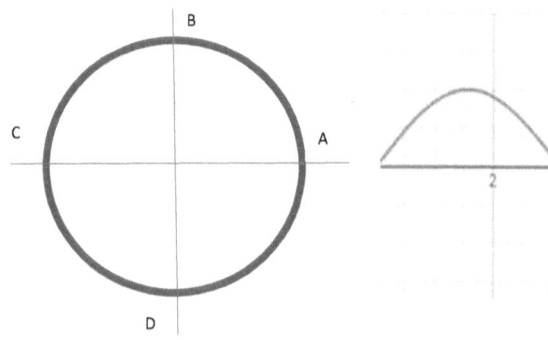

5. For the Pendulum below answer the following:

Label the Points of Maximum Potential Energy and Maximum Kinetic Energy:

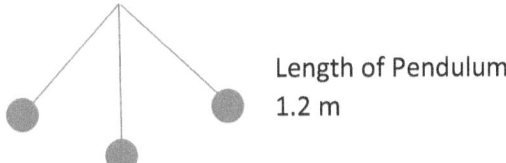

Length of Pendulum 1.2 m

A. With what Velocity will the Pendulum be moving at the very bottom, or equilibrium position? It was dropped from a height of 0.3m:

B. What is the Frequency of Oscillation if the Length of the String is 0.60m?

$$\omega = \sqrt{\frac{g}{L}}$$

M. What is the Angular Frequency of Oscillation?

C. Write the Equation for this Wave:

$$\psi = A \cos(2\pi f t + \theta)$$

A = amplitude
f = frequency
θ = initial displacement

D. Graph this Wave and state the equation:

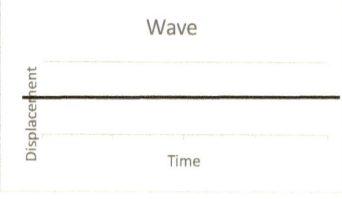

Equation:

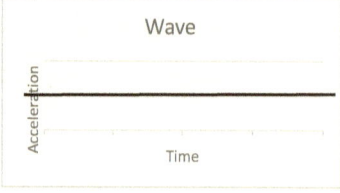

Equation:

Equation:

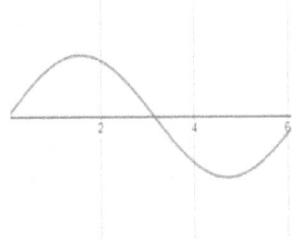

E. What Kind of Wave is Light:

F. What is the Speed of Light?

G. What is Light made of?

Electromagnetic Spectrum:

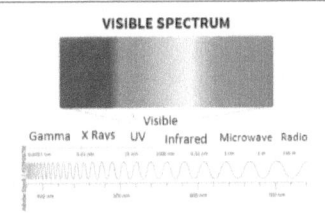

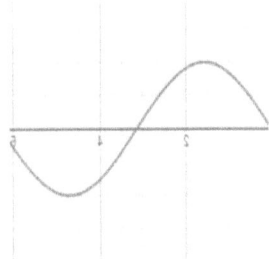

H. Give an example of Infra-Red Radiation:

I. Give an example of Ultraviolet Radiation:

J. List in order the colors of light from most energetic to least energetic:

K. List in order the colors of light from longest Wavelength to shortest Wavelength:

Name_____ Period_____

Waves 1

1…What is the Period of the Earth's Rotation around its axis?

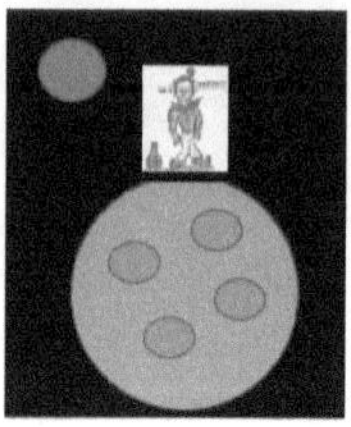

2…What is the Period of the Earth's Rotation around the sun?

3…If a Particle completes 6 revolutions per second around a circular path at a Radius of 3.0m, what is its

Frequency _____

Period_____

Speed_____

Centripetal Acceleration_____

4…What must be done to both the Amplitude and Frequency of a wave in order for it to have more Energy?

5...What must be done to both the Amplitude and Frequency of a wave in order for it to have less Energy?

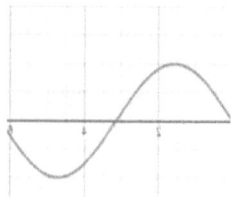

6...Give four examples of a Sinusoidal Wave Function:

7...In the figure below label the following: Amplitude, Wavelength, Crests, Troughs.

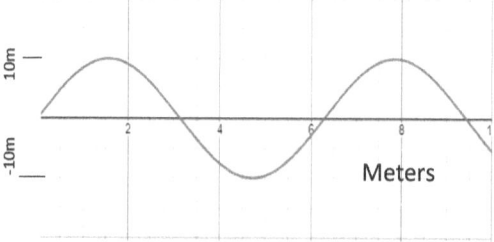

8...In the wave below, what is the Period and Frequency?

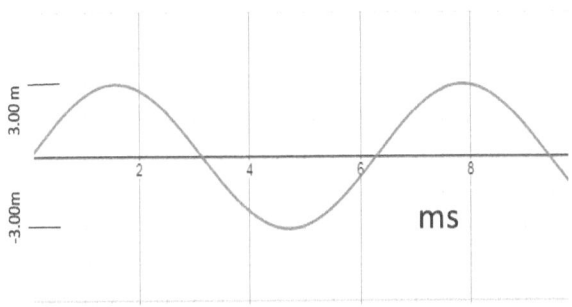

ms

9... In the wave below, what is the Period and Frequency?

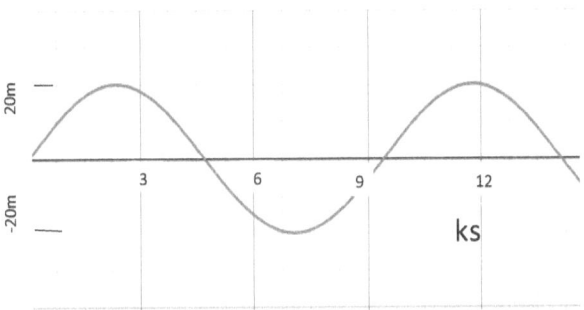

ks

10...If the wave in problem 9 has a Speed of 10m/s, what is its Wavelength?

11…What is the Wavelength of Sound of Frequency 340hz if its Speed at Standard Pressure and Temperature is 343 m/s?

12…What is the Period of the Sound in problem 11?

Waves Notes:

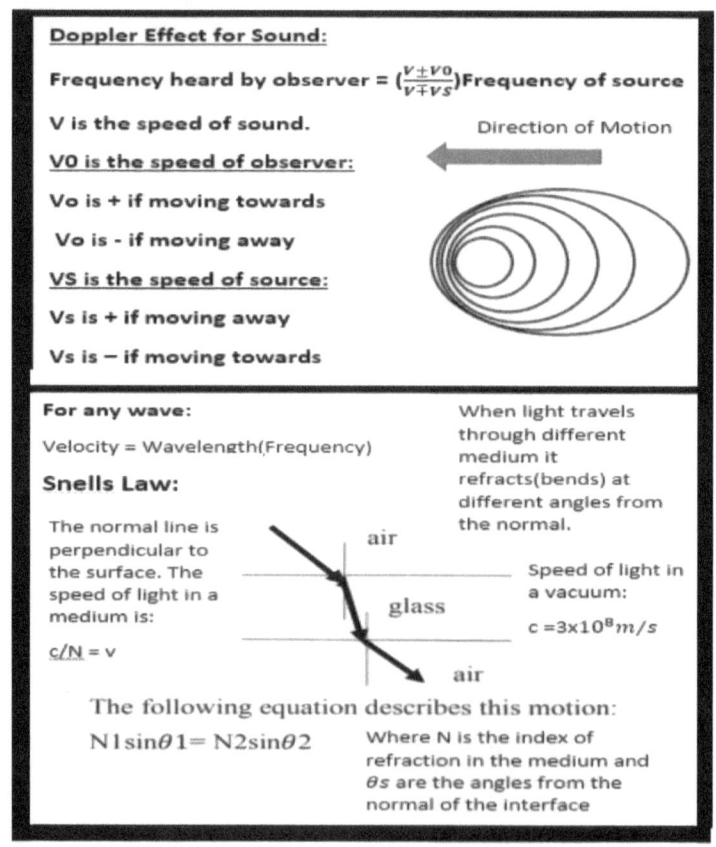

Doppler Effect for Sound:

Frequency heard by observer = $(\frac{V \pm VO}{V \mp VS})$ Frequency of source

V is the speed of sound.

VO is the speed of observer:

Vo is + if moving towards

Vo is − if moving away

VS is the speed of source:

Vs is + if moving away

Vs is − if moving towards

Direction of Motion

For any wave:

Velocity = Wavelength(Frequency)

Snells Law:

The normal line is perpendicular to the surface. The speed of light in a medium is:

$c/N = v$

When light travels through different medium it refracts(bends) at different angles from the normal.

Speed of light in a vacuum:

$c = 3 \times 10^8 m/s$

air

glass

air

The following equation describes this motion:

$N1 \sin\theta1 = N2 \sin\theta2$

Where N is the index of refraction in the medium and θs are the angles from the normal of the interface

Name_____Period____

Refraction and Reflection Review

Snell's Law Find the angle of refraction when light goes through air at the following angles with index of refraction 1.000 to the following materials with index of refractions:

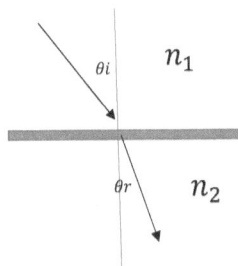

Index of Refraction	Incident Angle	Angle of Refraction	Speed of Light
1.1	20°		
1.5	26°		
1.9	30°		
2.0	28°		
2.3	10°		
2.8	18°		

Find the Critical Angle when light goes through a medium to air if the index of refraction of the glasses are:

Index of Refraction	Critical Angle
1.0	
1.2	
1.3	
1.4	
1.5	
1.6	
1.7	

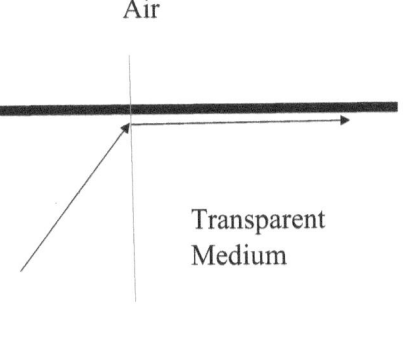

The angle of reflection is the same as the incident angle:

$$\theta i = \theta r$$

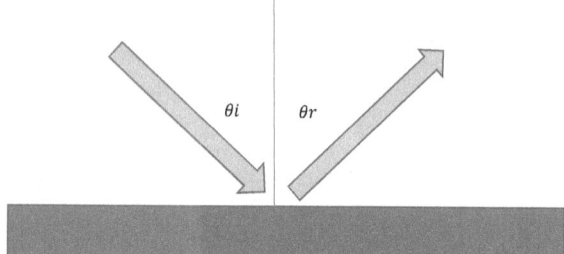

1-What is Critical Angle?

2-What is the Normal Line?

3-What happens when light is incident between two materials with an angle greater than the Critical Angle?

Name_____Period_____

Waves Packet 2

1...At 10.0m from the source of a series of waves the intensity is 1.00 w/m^2 and the Amplitude is 1.00 m.

State the Amplitude and Intensities at the following distances from the source:

Distance	Intensity	Amplitude
5.00 m		
15.0 m		
	2.00 w/m^2	
		3.00 m
30.0 m		
	0.50 w/m^2	
		0.50m

Source

2...Light ray is sent from a medium with Index of Refraction n= 2.30 to air n= 1.002. Find the following:

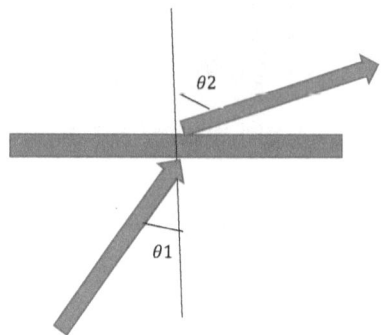

What is the angle of refraction if the incident angle is:

15°_____

20°_____

10°_____

What is the Critical Angle?

What is the speed of light at both mediums?

What will happen to the rays of light with incident angles larger than the Critical Angle?

3...Light ray is sent from a medium with Index of Refraction n= 3.30 to air n= 1.002. Find the following:

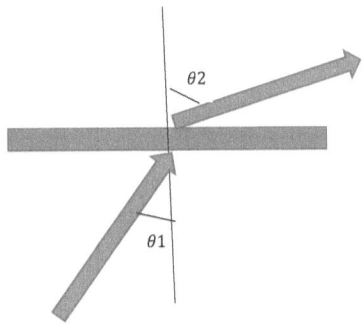

What is the angle of refraction if the incident angle is:

12°_____

20°_____

10°_____

What is the Critical Angle?

What is the speed of light at both mediums?

What will happen to the rays of light with incident angles larger than the Critical Angle?

4...Draw rays of light moving from a medium of higher Index of Refraction to a lower Index of Refraction:

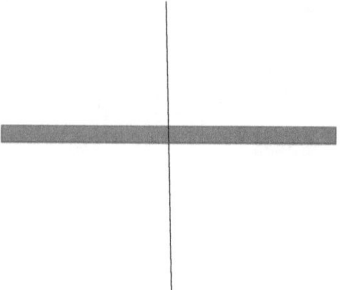

4...Draw rays of light moving from a medium of lower Index of Refraction to a higher Index of Refraction:

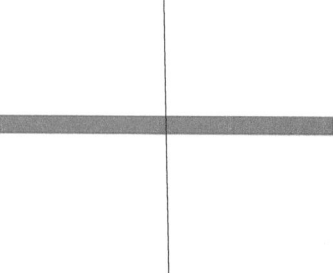

Name_____Period____

Waves, Intensity, and Interference:

1...The Intensity of a Sound Wave at a distance 4.00 m from the source is 1.00 W / m^2 and its Amplitude is 0.5 m.

A...What is the Intensity and Amplitude at the following Distances?

Distance	Intensity	Amplitude
1.00 m		
		10. m
3.00 m		
		1.0 m
6.00 m		
7.00 m		
	3.00 W / m^2	

2...When light goes through matter what happens to the following?

Frequency=

Period =

Wavelength =

Speed =

3...Light is incident on a transparent material from air with index of refraction of 1.56 at an angle of 30 degrees from the normal. Answer the following:

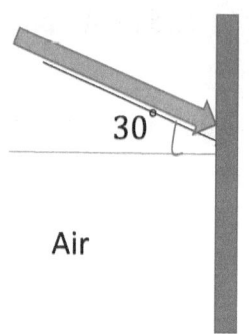

Air   Transparent material

A...What is the Speed of light in the medium?

B...What is the Angle of Refraction inside the medium?

C...Draw the ray of light inside the new medium.

4...Light goes from a transparent medium of index of refraction of 1.78 to air at an angle from the normal. Answer the following:

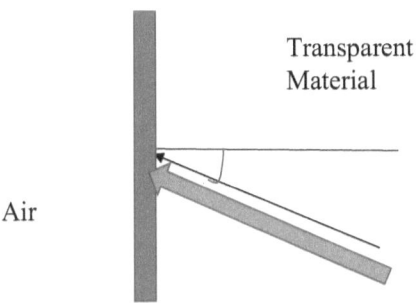

A...What is the definition of a Critical Angle?

B...What is the Critical Angle?

C...What is the Speed of light inside the transparent medium?

5...Light goes from a transparent medium of index of refraction of 1.28 to air at an angle of 20 degrees from the normal. Answer the following:

A...Draw the situation:

B...What the Speed of light in the transparent medium?

C...What is the Angle of Refraction of the light ray when it reaches air?

D...Draw the Ray of Refraction.

Name_____ Period_____

Waves Packet 3

1...At 5.0m from the source of a series of waves the intensity is 4.00 w/m^2 and the Amplitude is 6.00 m.

State the Amplitude and Intensities at the following distances from the source:

Distance	Intensity	Amplitude
7.00 m		
15.0 m		
	2.00 w/m^2	
		3.00 m
30.0 m		
	0.50 w/m^2	
		0.50 m

Source

2...Light ray is sent from a medium with Index of Refraction n= 2.70 to air n= 1.002. Find the following:

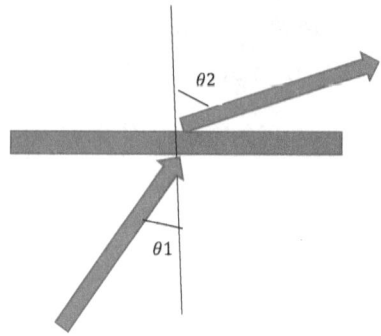

What is the angle of refraction if the incident angle is:

15° _____

9° _____

10° _____

What is the Critical Angle?

What is the speed of light at both mediums?

What will happen to the rays of light with incident angles larger than the Critical Angle?

3...Light ray is sent from a medium with Index of Refraction n= 4.00 to air n= 1.002. Find the following:

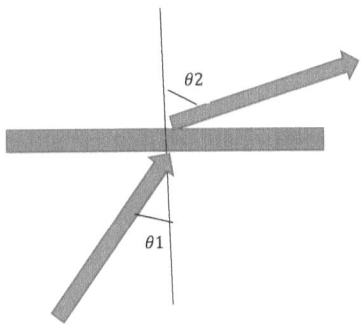

What is the angle of refraction if the incident angle is:

12°_____

9°_____

10°_____

What is the Critical Angle?

What is the speed of light at both mediums?

What will happen to the rays of light with incident angles larger than the Critical Angle?

4...Draw rays of light moving from a medium of higher Index of Refraction to a lower Index of Refraction:

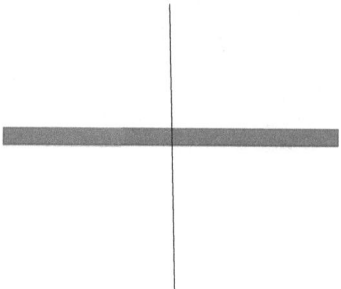

4...Draw rays of light moving from a medium of lower Index of Refraction to a higher Index of Refraction:

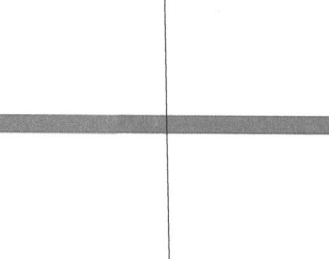

Name_____Period____

Refraction and Reflection Review

Snell's Law

Find the angle of refraction when light goes through air at the following angles with index of refraction 1.000 to the following materials with index of refractions:

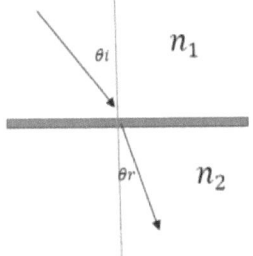

Index of Refraction	Incident Angle	Angle of Refraction	Speed of Light
1.3	10°		
1.4	16°		
1.5	10°		
2.7	18°		
2.8	10°		
2.9	12°		

Find the Critical Angle when light goes through a medium to air if the index of refraction of the glasses are:

Index of Refraction	Critical Angle
2.0	
2.2	
2.3	
2.4	
2.5	
2.6	
2.7	

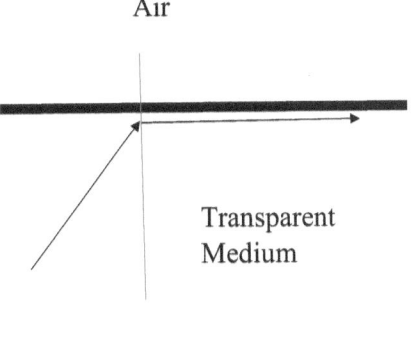

Conceptual Questions

1... When the brakes of a car are applied what happens with the Internal Energy of the Brakes? Will there be a rise or drop of its Temperature?

2... When two blocks at different Temperature are in contact with each other, which block loses Energy and which block gains Energy?

$-C_2 m \Delta t = C_1 m \Delta t$ both of their Final Temperature will be the same

3... A Pressure Volume Graph would be a curve but it can be made linear if Pressure is graphed with the reciprocal of Volume:

$$P = \frac{nRT}{1}\left(\frac{1}{V}\right)$$

In the Equation above what is the Slope of the Graph?

4...What happens to the Intensity of light with distance from Source?

Intensity $\propto \frac{1}{r^2}$

5...What is the Angle of Refraction when the Incident Angle is the Critical Angle?

6...Standing Waves have Nodes and Antinodes. One Wavelength has two Antinodes and three Nodes:

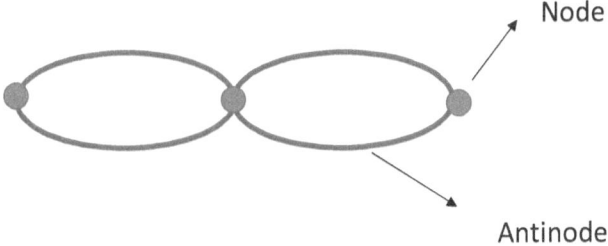

7...The Net Force is the sum of the Forces.

8...The Atwood Machine is an Ideal pulley that is Frictionless and Massless.

9...When two objects collide in Closed System and stick what is conserved and what is lost?

10...What is the Area of a Circle if the Uncertainty of its Radius is a given amount? What is the Uncertainty of the Area if multiplication is involved?

11...How much farther does something drop at the second Second, compared with the first Second?

12...How can you calculate Moles and Number of Gas Particles by knowing its Volume, Pressure, and Temperature?

PV = nRT   should Temperature be in Celsius or Kelvins?

If a gas expands what happens to its density?

What happens to Energy when Temperature rises?

14...How can you calculate the Frequency of a Wave by knowing its Period?

What is the Period of a Spring Mass System Equation?

Would it be $T = 2\pi \sqrt{\frac{m}{k}}$?

What is the relationship between the hanging mass on a Spring and the Spring Constant?

What is the Amplitude of a Spring Mass System if the Amplitude is Maximum Displacement?

Restoring Force = -k(Amplitude)

Where the Restoring Force is equal to the Weight of the Hanging Mass = mg

Elastic Potential Energy is = $(1/2)k\Delta x^2$

Speed of a Wave = (Wavelength)(Frequency)

Name_____Period_____

Double Slit, Oscillations, Sound

Below are Standing Waves in a Guitar String that is 0.80 m long. The Speed of the Wave is 60 m/s. Fill the table below:

Harmonic	Wavelength	Frequency

Harmonic	Number of Nodes	Number of Anti Nodes

Fill the following for Sound Waves in a Pipe:

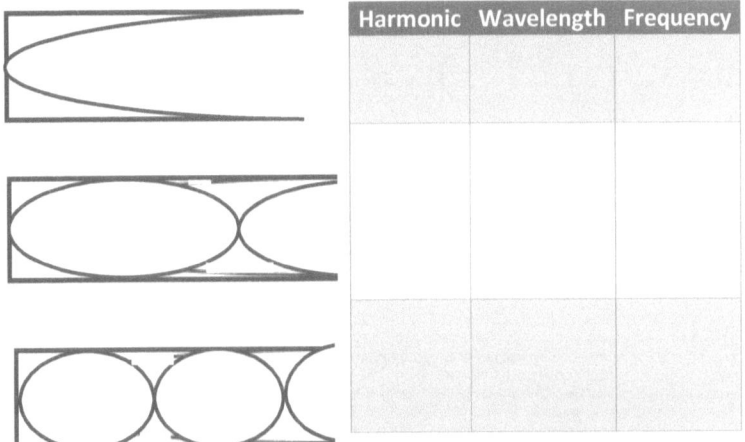

Harmonic	Wavelength	Frequency
1st	8.00 m	42.9 Hz
3rd	2.67 m	128.6 Hz
5th	1.60 m	214.4 Hz

Velocity of Sound in Standard Conditions is 343 m/s. The Length of the Pipe is 2.00 m:

Harmonic	Number of Nodes	Number of Anti Nodes
1st	1	1
3rd	2	2
5th	3	3
7th	4	4

Pipe open at one end

Fill the following for Sound Waves in a Pipe:

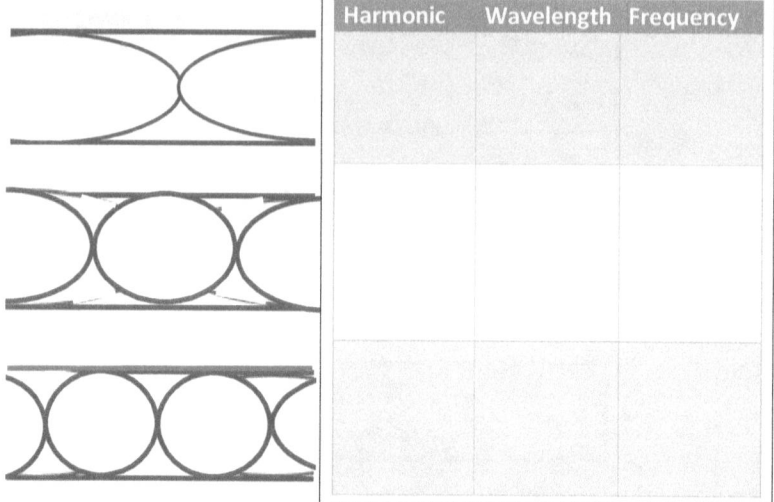

Harmonic	Wavelength	Frequency

Velocity of Sound in Standard Conditions is 343 m/s. The Length of the Pipe is 2.00 m:

Harmonic	Number of Nodes	Number of Anti Nodes

Pipe open at both ends

Energy:

Kinetic Energy = Energy of Motion

$KE = (1/2)mv^2$

Potential Energy = Energy Stored

PE for gravity near Earth's Surface = mgh

PE for spring = $(1/2)kx^2$ where x is the amount that it is stretched or compressed.

Spring:

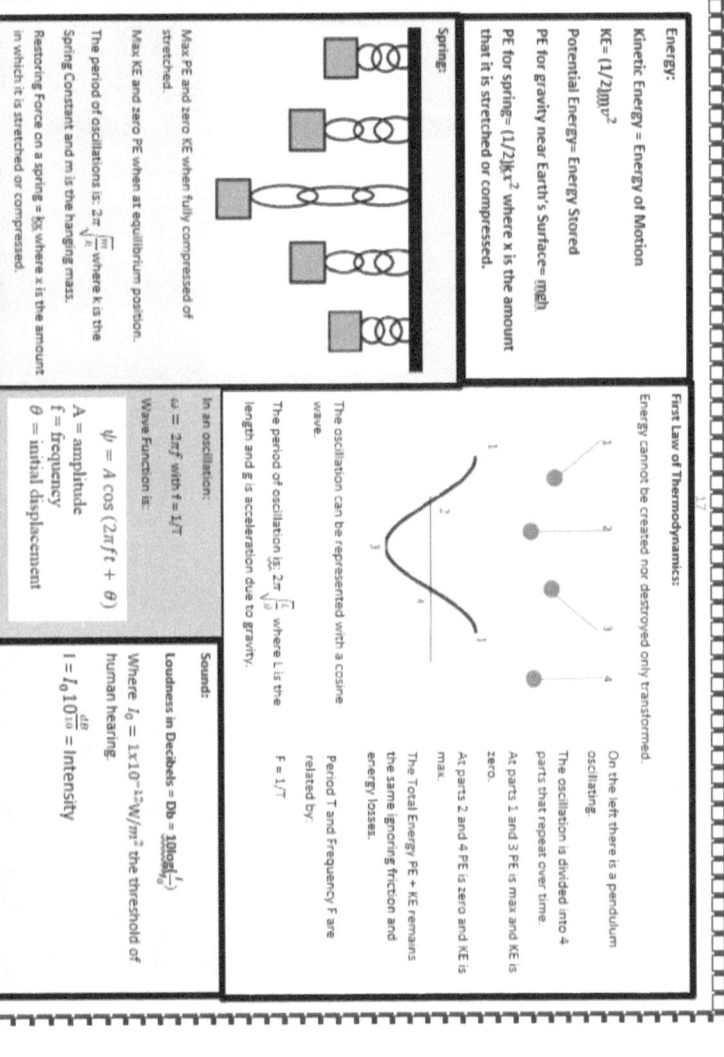

Max PE and zero KE when fully compressed of stretched.

Max KE and zero PE when at equilibrium position.

The period of oscillation is: $2\pi\sqrt{\frac{m}{k}}$ where x is the Spring Constant and m is the hanging mass.

Restoring Force on a spring = kx where x is the amount in which it is stretched or compressed.

In an oscillation:

$\omega = 2\pi f$ with $f = 1/T$

Wave Function is:

$$\psi = A\cos(2\pi f t + \theta)$$

$A = $ amplitude
$f = $ frequency
$\theta = $ initial displacement

First Law of Thermodynamics:

Energy cannot be created nor destroyed only transformed.

On the left there is a pendulum oscillating.

The oscillation is divided into 4 parts that repeat over time.

At parts 1 and 3 PE is max and KE is zero.

At parts 2 and 4 PE is zero and KE is max.

The Total Energy PE + KE remains the same ignoring friction and energy losses.

Period T and Frequency F are related by:

$F = 1/T$

The oscillation can be represented with a cosine wave.

The period of oscillation is: $2\pi\sqrt{\frac{L}{g}}$ where L is the length and g is acceleration due to gravity.

Sound:

Loudness in Decibels = $Db = \frac{10 \log(\frac{I}{I_0})}$

Where $I_0 = 1 \times 10^{-12} W/m^2$ the threshold of human hearing.

$I = I_0 10^{\frac{dB}{10}} = $ Intensity

Double Slit Interference Explanation:

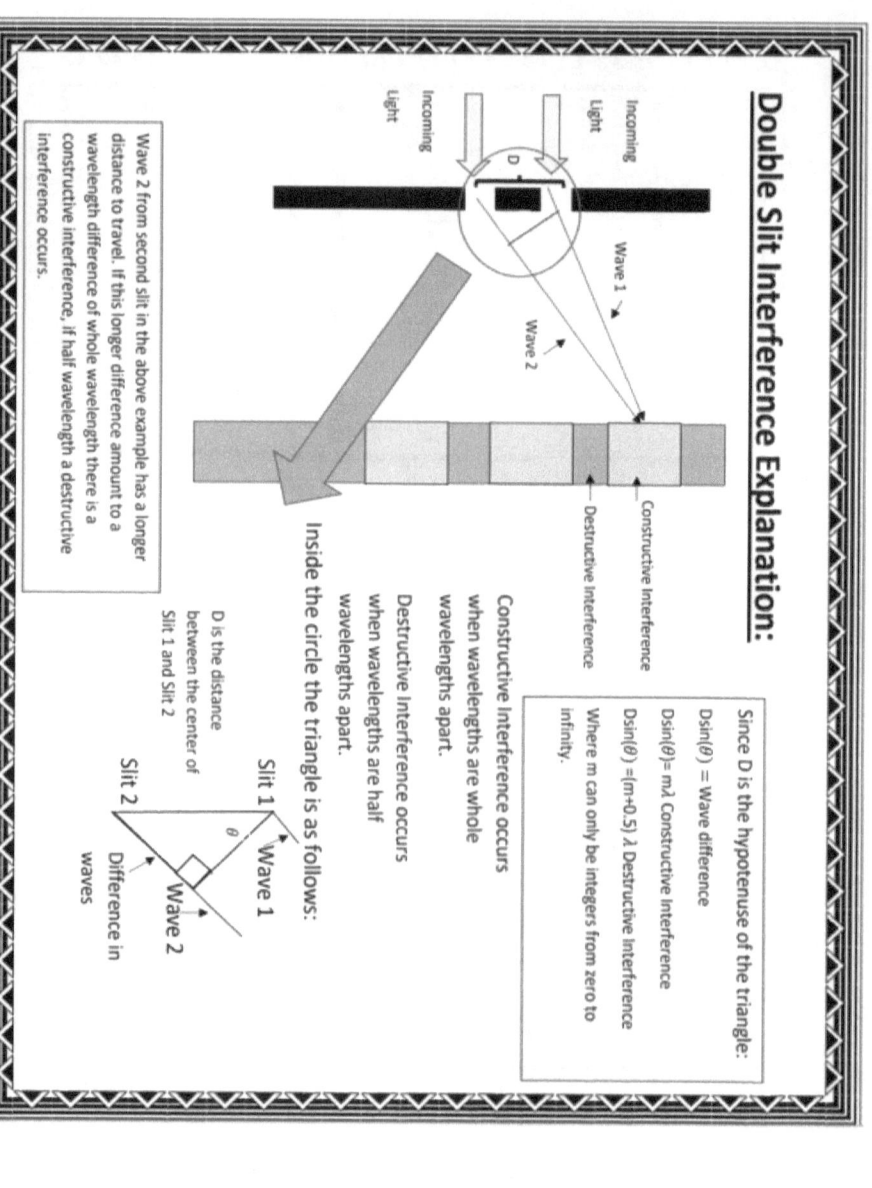

Since D is the hypotenuse of the triangle:

$D\sin(\theta)$ = Wave difference

$D\sin(\theta) = m\lambda$ Constructive Interference

$D\sin(\theta) = (m+0.5)\lambda$ Destructive Interference

Where m can only be integers from zero to infinity.

Constructive Interference occurs when wavelengths are whole wavelengths apart.

Destructive Interference occurs when wavelengths are half wavelengths apart.

Inside the circle the triangle is as follows:

D is the distance between the center of Slit 1 and Slit 2

Wave 2 from second slit in the above example has a longer distance to travel. If this longer difference amount to a wavelength difference of whole wavelength there is a constructive interference, if half wavelength a destructive interference occurs.

Doule slit experiment

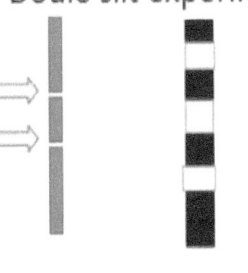

Interference occurs with a pattern of bright and dark spots on the screen after light goes through the two slits.

Light waves interaction leads to interference patterns

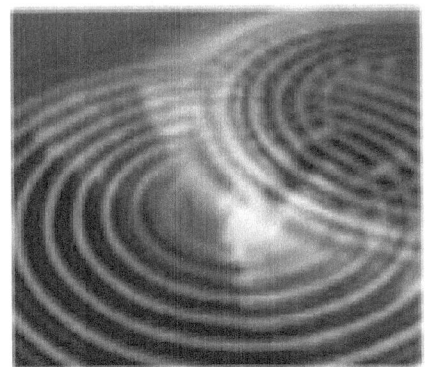

Light is an Electromagnetic Radiation which can be seen as both a particle and as a wave depending on the condition. It is called the wave-particle duality.

For the Double Slit Experiment, answer the following:

Equation:

$$\frac{\lambda D}{d} = s$$

Where D is the distance to the screen, λ is Wavelength, s is the fringe spacing, and d is the distance between the Slits.

Explain what happens to the spacing of fringes if:

Change	Result
Screen brought closer	
Screen brought farther	
Slit spacing increased	
Slit spacing decreased	
Wavelength of light increased	
Wavelength of light decreased	

Label the parts of the fringe pattern below:

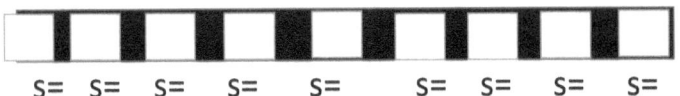

S=   S=   S=   S=   S=   S=   S=   S=   S=

Calculate the fringe spacing for red light (choose the wavelength for this red light_____nm) when the screen is 3.00m away, and the slit separating is 1.00 mm:

What is the phase shift between two wavelengths that leads to a destructive interference?

What is the phase shift between two wavelengths that leads to a constructive interference?

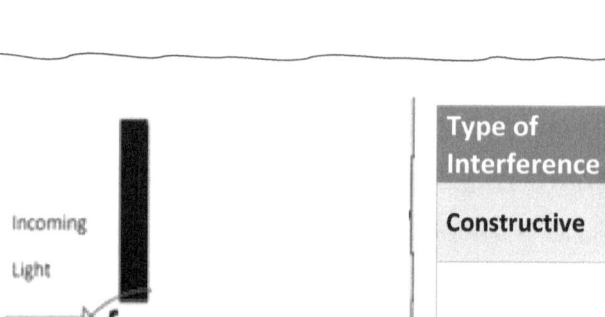

Type of Interference	Phase Shift
Constructive	

Use the red light from previous page to complete the table above

Label the parts of the fringe pattern below:

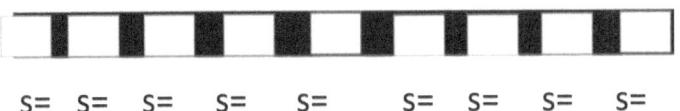

S=   S=   S=   S=   S=   S=   S=   S=   S=

Calculate the fringe spacing for blue light (choose the wavelength for this blue light_____nm) when the screen is 3.00m away, and the slit separating is 1.00 mm:

What is the phase shift between two wavelengths that leads to a destructive interference?

What is the phase shift between two wavelengths that leads to a constructive interference?

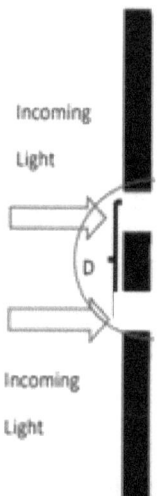

Type of Interference	Phase Shift
Constructive	

Use the blue light from previous page to complete the table above

Name_____Period__

Sound

1- What is the Intensity of a Sound of 30dB?

2- What is the Intensity of Sound at 40dB?

3- How many times is the Intensity 40dB bigger than the 20dB Sound?

4- How many times is the Intensity of 100dB bigger than 80 dB?

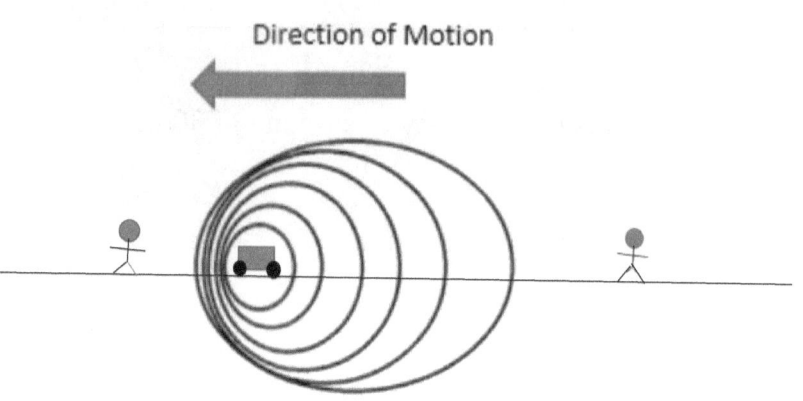

5- A source of Sound is moving towards an observer. What can you say about the Frequency and Wavelength of Sound? (bigger or smaller than the actual)?

6- A source of Sound is moving away from an observer. What can you say about the Frequency and Wavelength of sound?

(bigger or smaller than the actual)?

7-Does Sound travel faster or slower when approaching an observer?

8-Does Sound travel faster or slower when moving away from an observer?

9-What is the Wavelength of a 345 m/s sound with Frequency of 90Hz?

10- How many decibels of Sound is produced with an Intensity that is 1000 times the threshold intensity?

11-Instead of Crests and Troughs, Sound Waves have what components?

12- How do Particles of a medium vibrate with respect to the propagation of Sound?

13...How loud is sound in the following intensities?

A...$2.2 \times 10^{-8} W/m^2 =$

B...$3.8 \times 10^{-2} W/m^2 =$

C...$4.6 \times 10^{-6} W/m^2 =$

D...$5.2 \times 10^{-11} W/m^2 =$

E...$6.2 \times 10^{-9} W/m^2 =$

14…What kind of wave is Sound?

15…What is the Intensity of the Sound at the following decibels?

A…40 dB

B…50 dB

C…77 dB

D…99 dB

16....

A- If the observer moves 30m/s towards a source of a Sound of 70 Hz that propagates at 345 m/s, what is the Frequency heard?

B- If the observer moves 20m/s away from a source of a Sound of 70 Hz that propagates at 345 m/s, what is the Frequency heard?

C- If the source moves 5m/s towards the observer with a Sound of 70 Hz that propagates at 345 m/s, what is the Frequency heard?

D- If the source moves 40m/s away from the observer with a Sound of 70 Hz that propagates at 345 m/s, what is the frequency heard?

E...If the person in problem D is now also moving towards the source of Sound at a speed of 5.0 m/s, what is the Frequency heard?

17-If you a double the Distance from a source of Sound, by how much does the Intensity decreases?

18-If you increase the Amplitude of the Sound 4 times, by how much has the Intensity increased?

19...If the Intensity of Sound at a Distance of 55 m is $3.2 \times 10^{-7} W/m^2$, what is the Intensity at the following Distances?

A...90.0m =

B...22.5m =

C...27.5 m =

D...165 m =

E...15.0 m =

Name_____Period___

Loudness

For each Decibel find its Intensity and how many times Threshold Hearing it is:

Decibel	Intensity	How many times threshold
30		
40		
50		
60		
70		
80		
90		
100		
110		
120		
130		
140		

Gravity and Electromagnetism

Objects with mass attract each other obeying the following equation:

$$\text{Force} = \frac{GM1M2}{r^2}$$

G is the Universal Gravitational Constant:

$6.67 \times 10\text{\^{}}-11 \ m^3/(kg \ s)$

M1 and M2 are the two masses in Kilograms and r is the Distance between them in Meters.

Charged objects can attract or repel each other obeying the following equation:

$$\text{Force} = \frac{Kq1q2}{r^2}$$

K is the Coulomb's Constant:

$9 \text{X} 10\text{\^{}}9 \ (Nm\text{\^{}}2) / (C\text{\^{}}2)$

q1 and q2 are the two charges in Coulombs and r is the Distance between them in Meters.

Like charges repel and unlike charges attract.

The equation for the Electric Field from a point charge is:

$$\text{Field} = \frac{Kq1}{r^2}$$

In the beginning it was the Big Bang from which the universe was formed:

After the Big Bang the grand cosmic unified force separated in to the four forces present in the universe today. These forces are:

1) **Gravitational:** Causing masses to attract other masses.
2) **Electromagnetic:** Giving charge to particles.
3) **Weak:** Breaks the nucleus of atoms.
4) **Strong:** Keeps the nucleus of atoms together.

Everything began at Big Bang. All things have a single common origin!

Electric Force between two particles:

$F = kQq/r^2$

$k = 9 \times 10^9 Nm^2/C^2 =$ Coulomb's Constant

Q and q being the two Charges

r being the Distance between them

Electric Field Lines are a field of photons emanating from the Particles. The Field Lines point towards the Negative Charges and away from the Positive charges.

These are the Electric Field Lines between two opposite charged Particles. The Force between them is attractive:

These are the Electric Field Lines between two like charged Particles. The Force between them is repulsive:

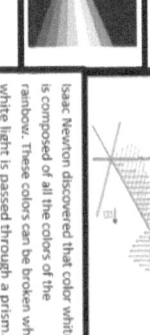

Light is an Electromagnetic Wave: The Electric and Magnetic Fields are always 90 degrees from each other.

Isaac Newton discovered that color white is composed of all the colors of the rainbow. These colors can be broken when white light is passed through a prism.

Name_____ Period____

Forces, Gravitation, and Electricity

1...Two boxes are connected by a string and are being pulled to the right with a Force of 30.0N. Calculate the following: Assume surface is frictionless.

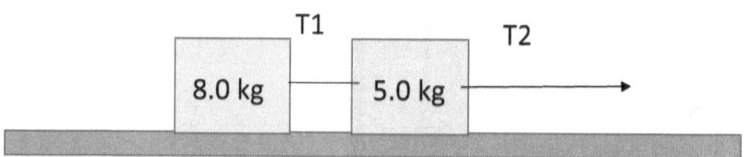

A...Net Force on the System:

B...Acceleration of the System:

C...Tension 1:

D...Tension 2:

2...Three boxes are connected by strings and are being pulled to the right with a Force of 80.0N. Calculate the following: Assume surface is frictionless.

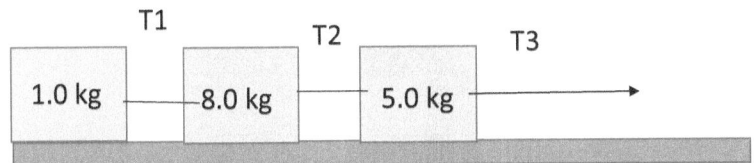

A...Net Force on the System:

B...Acceleration of the System:

C...Tension 1:

D...Tension 2:

E...Tension 3:

3...A Spring with Spring Constant K= 8.0 N/m has a length of 4.0 m. A 3.00 kg box is about to be hanged of the spring:

A....When the block hangs on the Spring, what is the amount of stretch?

B...What is the Period if it begins to oscillate?

C...What is the Frequency of Oscillation?

D...What is the Angular Frequency?

4...A ball of mass 6.00 kg is at a distance of 1.0 cm from a ball of mass 7.00kg.

A...What is the Gravitation Force between them? Is it attractive of repulsive?

B...What will be the new Force if the distance is cut by a half and one of the masses tripled?

5...A ball with 9.0μC and another of 10 μC are placed 6.0mm apart.

A...What is the Electric Force between them? Is it attractive of repulsive?

B...What will be the new Force if the distance is cut by a third and one of the masses decreased to a third?

6...Below are Standing Waves in a Guitar String that is 1.00 m long. The Speed of the Wave is 50 m/s. Fill the table below:

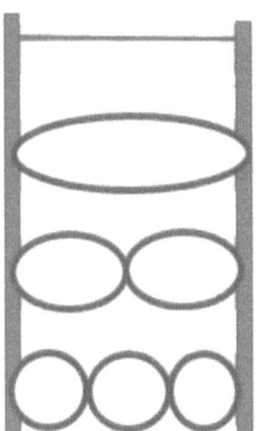

Harmonic	Wavelength	Frequency

Harmonic	Number of Nodes	Number of Anti Nodes

7...What is the Period of the Earth's Rotation around its axis?

8...What is the Period of the Earth's Rotation around the sun?

9...If a Particle completes 8 revolutions per second around a circular path at a Radius of 2.0m, what is its

Frequency _____

Period _____

Speed_____

Centripetal Acceleration_____

10...Two metal spheres one with a Charge of $+2.3 \times 10^{-5} C$ and the other with a Charge of $-3.3 \times 10^{-5} C$ are separated by a Distance of 2.0 cm. Answer the following:

A...What is the Force between the two spheres?

B...Is the Force repulsive or attractive?

C...What is the new Force if the distance between the two is cut in half?

Electric Forces

1- What is the Electric Force between a Proton and an Electron 1.00 cm apart?

2- What is the Gravitational Force between the Proton and the Electron 1.00 cm apart?

3-What is the Electric Force on a 2.00 kg Particle of Charge $3.00\mu C$ due to two other Particles with the same mass separated by a Distance of 3.0 cm and charges as shown below:

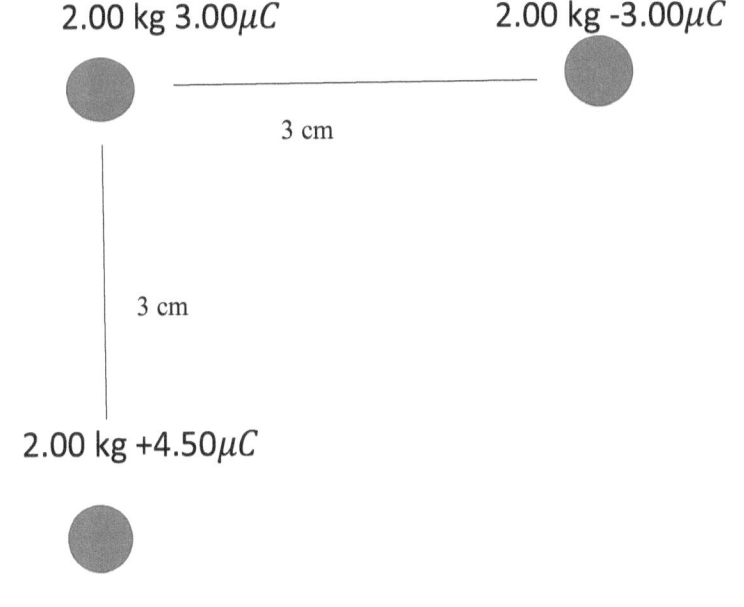

Find both Magnitude of the Net Force and Direction:

Use the following table to answer the question:

X	Y
Sum:	**Sum:**

Draw the Net Force Vector on the 2.00kg Particle with Charge $3.00\mu C$ using a right triangle:

4-Now do the same thing on problem 3 but for the Force of Gravity:

Use the following table to answer the question:

X	Y
Sum:	**Sum:**

Draw the Net Force Vector on the 2.00kg Particle with mass 2.00 kg using a right triangle:

5- What is the Electric Force on a 1.00 kg Particle of Charge 7.00μC due to the two other Particles:

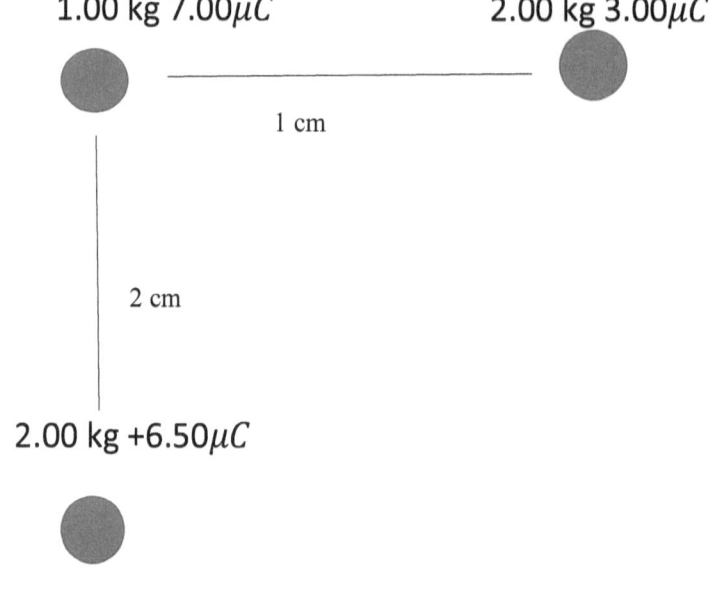

1.00 kg 7.00μC

2.00 kg 3.00μC

1 cm

2 cm

2.00 kg +6.50μC

Find both Magnitude of the Net Force and Direction:

Use the following table to answer the question:

X	Y
Sum:	**Sum:**

Draw the Net Force Vector on the 1.00kg Particle with Charge $7.00\mu C$ using a right triangle:

6- What is the Electric Force on a 3.00 kg Particle of Charge $4.00\mu C$ due to the two other Particles:

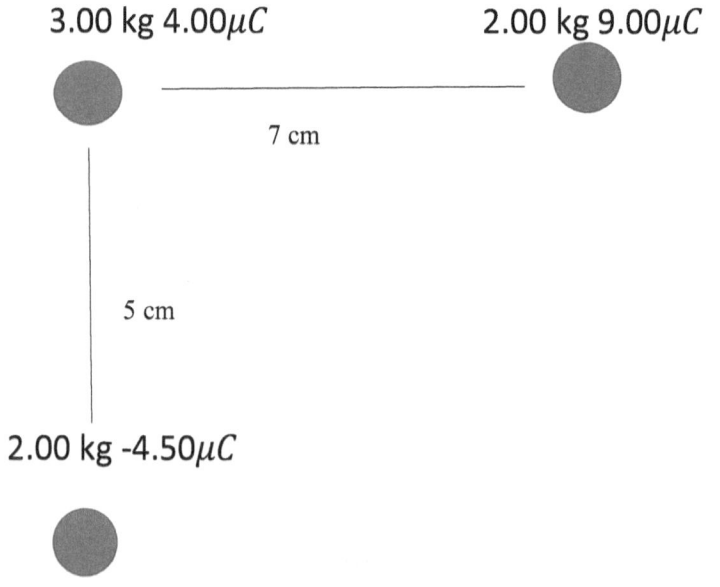

Find both Magnitude of the Net Force and Direction:

Use the following table to answer the question:

X	Y
Sum:	**Sum:**

Draw the Net Force Vector on the 3.00kg Particle with Charge 4.00μC using a right triangle:

7-What is the Electric Force on a 5.00 kg Particle of Charge $2.00 \mu C$ due to the two other Particles:

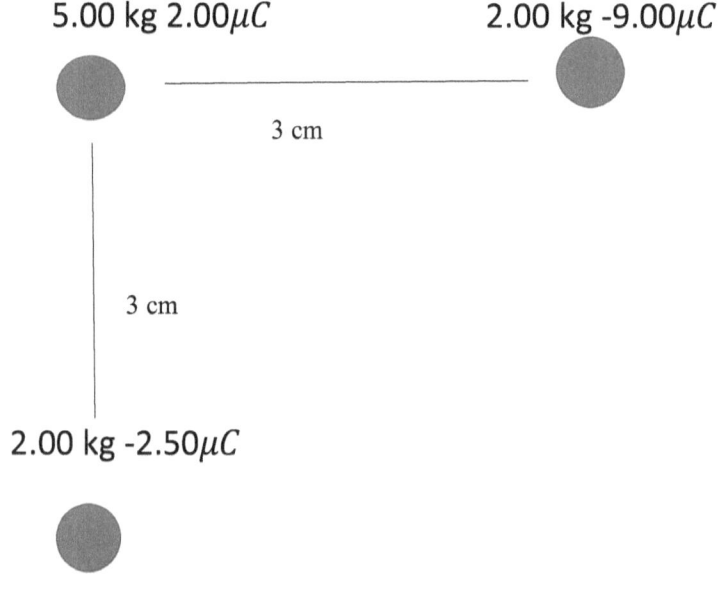

Find both Magnitude of the Net Force and Direction:

Use the following table to answer the question:

X	Y
Sum:	**Sum:**

Draw the Net Force Vector on the 5.00kg Particle with Charge $2.00\mu C$ using a right triangle:

Name_____ Period_____

Electric Forces 2

1...Two metal spheres one with a Charge of $+1.3 \times 10^{-5} C$ and the other with a Charge of $-2.3 \times 10^{-5} C$ are separated by a Distance of 3.0 cm. Answer the following: **(In the following problems always use the result of the Force in Problem A)**

A...What is the Force between the two spheres?

B...Is the Force repulsive or attractive?

C...What is the new Force if the Distance between the two is cut in half?

D…What is the new Force if the Charge on both is tripled?

E…What is the new Force if the Charge on only one Charge is tripled?

F…What is the new Distance between the Particles if their new Force is five times greater?

G...By how much must you increase the Distance between the two Particles for the Force to decrease ten times?

2...A Proton and an Electron are separated by a distance of $7.0 \times 10^{-8} m$.

A...What is the Electric Force between them?

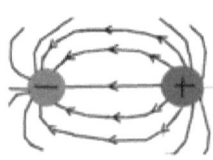

B...What is the Gravitational Force between them?

C…How much greater is the Electric Force compared with the Gravitational Force?

3…What is the Charge of two equal Particles if they are 1.00 m apart, and their Force is 1.00 N?

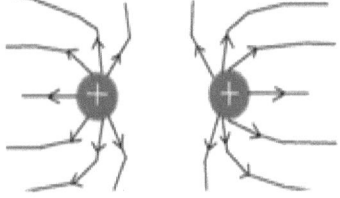

4... What is the Charge of two Particles if the Charge in one is three times greater than the other, both are negative and their separation is 3.00 m, and the Force between them is $7.8 \times 10^{-3} N$?

5... How many Electrons must be removed from a conducting sphere to give it a Charge of +9.0 C?

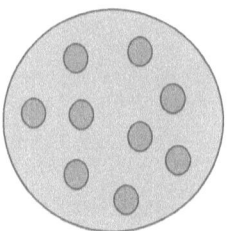

Name_____Period____

Electricity and Gravity Review

1...Find the Force of Gravity between masses 7.00 g and 7.00 g at a distance of 7.00 m apart:

2...If the Distance between objects is tripled, by how much does the Force change?

3...If one mass is halved by how much does the Force change?

4....If one mass triples and the other drops to a 1/3 by how much does the Force change?

5...If the distance is tripled and one mass is also tripled by how much does the Force change?

6...If one mass is cut in half and the distance is tripled, by how much does the Force change?

7...What is the Electric Force between a -7.00C and a 7.00C particle a distance of 7.00 m apart?

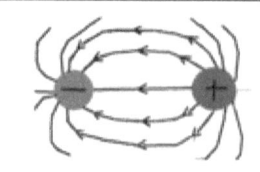

8...Is the Force Attractive or Repulsive?

9...What is the Distance between two 8.00C particles if their repulsion is 100. N?

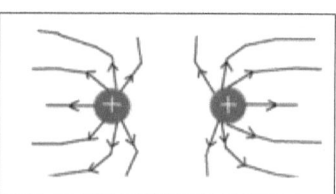

10... Find the Net Electric Force on the star labeled A.

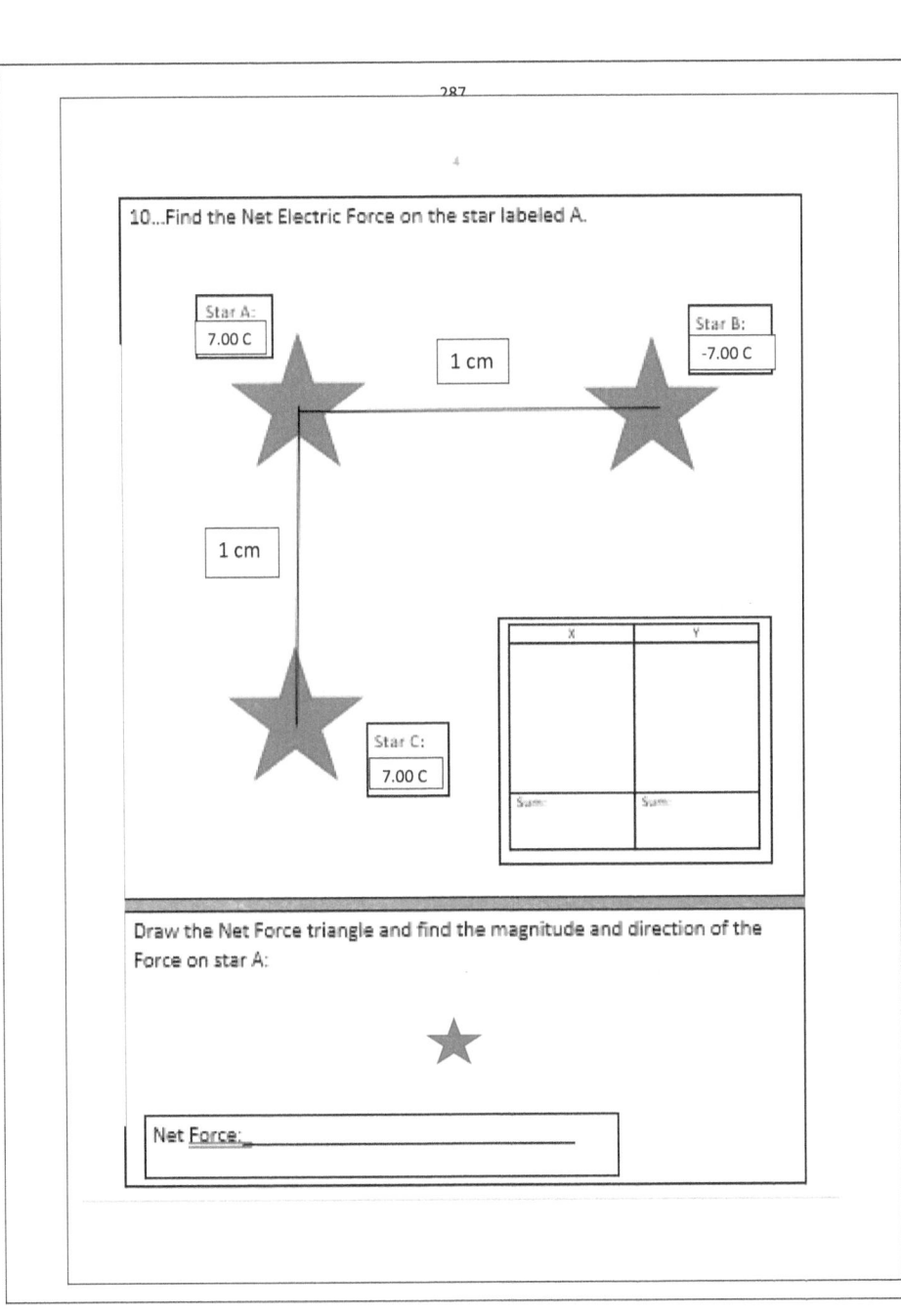

Draw the Net Force triangle and find the magnitude and direction of the Force on star A:

Net Force: _____

11....What are the 4 Forces of nature?

    1:                              3:

    2:                              4:

12...Electric Field lines point_____ Positive Charge and _____ Negative Charge.

13...Draw Electric Field Lines around the particles:

Charging Objects:

An object can become charged when it acquires an excess of Positive Holes or Electrons with a Negative Charge.

Positive Holes exist in the Electron Shells in Atoms with missing Electrons. When an Atom loses an Electron, it gains a Vacant Spot called a Hole which is positive.

    Vacant Spot   Filled Spot
    Positive Hole   Negative Charge

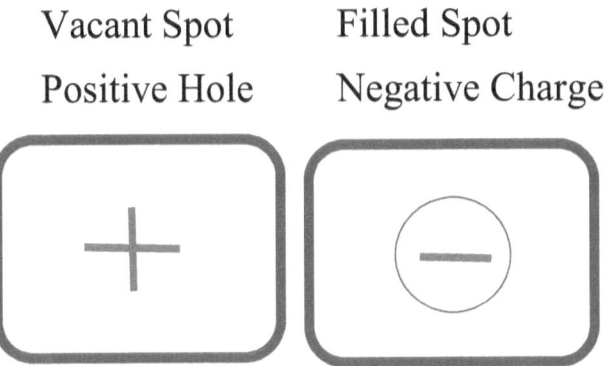

If an Atom has more Electrons than Protons, it has a Negative Charge. If the Number of Electrons equals to the number of Protons the Atom is Neutral. If there are Less Electrons than

Protons, the Atom has Positive Holes, and it has a Positive Charge.

There are three ways for materials to gain charge:

1...**Charging by Friction.** Rubbing your hair with a Styrofoam and seeing the Static Force between your hair and the Styrofoam. During the rubbing process the Styrofoam loses Electrons, while your hair gains Electrons. The result is that since unlike Charges Attract, your hair is attracted by the Styrofoam.

Just like Energy cannot be created nor destroyed only transformed, the number of Charges is held constant. At first the hair and Styrofoam are Neutral.

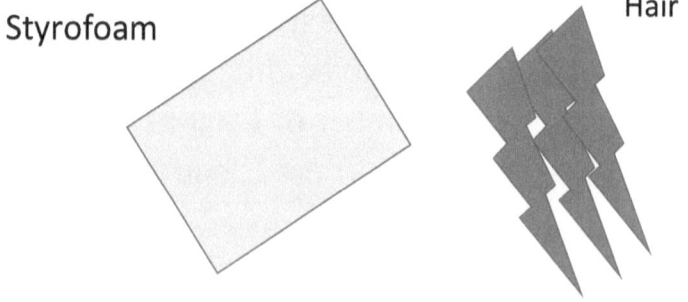

After the rubbing process Electrons jump from Styrofoam to the hair. The Net Charge on the System however is still zero, but now the charges are separated between the hair and Styrofoam.

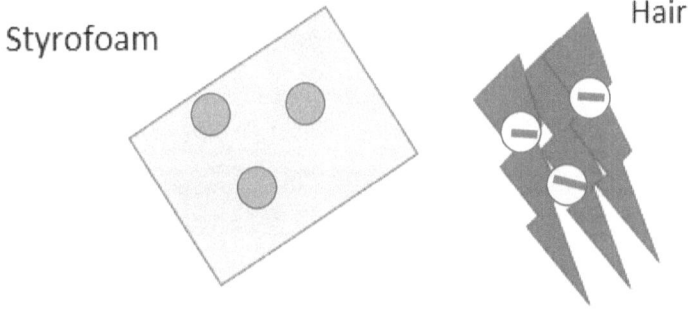

Suppose that the Styrofoam now has 3 Holes and the hair 3 Electrons. The Net Charge on the System is still zero: 3-3 = 0, but the two objects are charged.

2...**Charging by Conduction.** When you bring the Charged Rod that touches another rod, charges flow and spread on both.

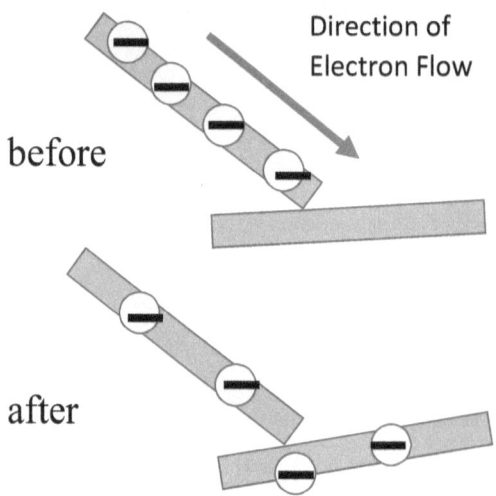

The Net Charge is kept the same at -4e obeying the Conservation Law. Charges like to spread on the materials that they are allowed to flow.

3…**Charging by Induction.** When you bring a Charge Rod near another rod **without touching.**

during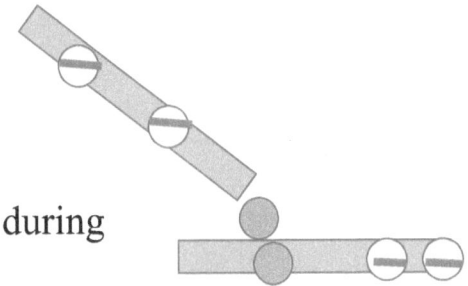

The Positive Holes of the Second Rod are attracted by the Negative Rod, and the Electrons of the Second Rod move away as most they can.

If the Second Rod is connected to the ground the Electrons will escape the Second Rod towards the ground leaving just the Positive Holes.

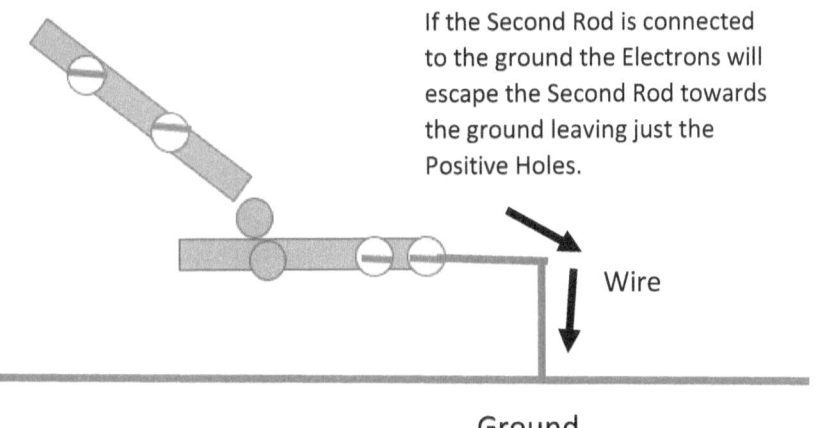

At the end we have a Negatively Charged Rod and a Positively Charged Rod.

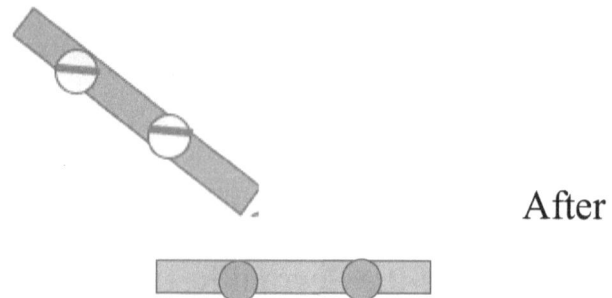

After

There are Three Types of Materials:

1...**Conductors** which allows Charges to flow freely.

2...**Insulators** which does not allow Charges to flow easily.

3...**Semiconductors** which facilitate Charges to flow but not as much as Conductors.

Name_____ Period_____

Electric Forces 3

1…Two metal spheres one with a Charge of $+2.3 \times 10^{-5} C$ and the other with a Charge of $-3.3 \times 10^{-5} C$ are separated by a Distance of 7.0 cm. Answer the following: **(In the following problems always use the result of the Force in Problem A)**

A…What is the Force between the two spheres?

B…Is the Force repulsive or attractive?

C…What is the new Force if the Distance between the two is cut to a third?

D...What is the new Force if the Charge on both is quadrupled?

E...What is the new Force if the Charge on only one Charge is halved?

F...What is the new Distance between the Particles if their new Force is three times greater?

G...By how much must you increase the Distance between the two Particles for the Force to decrease eight times?

2...A Proton and an Electron are separated by a distance of $3.0 \times 10^{-8} m$.

A...What is the Electric Force between them?

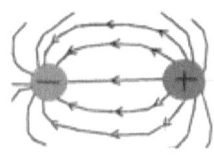

B...What is the Gravitational Force between them?

C...How much greater is the Electric Force compared with the Gravitational Force?

3...What is the Charge of two equal Particles if they are 2.00 m apart, and their Force is 3.00 N?

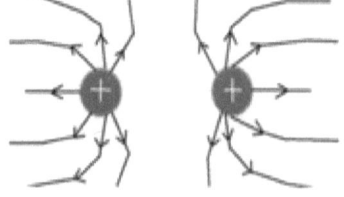

4...What is the Charge of two Particles if the Charge in one is four times greater than the other, both are negative and their separation is 1.00 m, and the Force between them is $2.8 \times 10^{-3} N$?

5...How many Electrons must be removed from a conducting sphere to give it a Charge of +1.0 C?

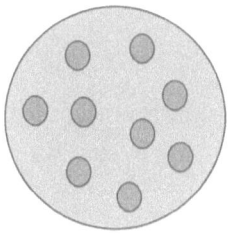

Name_____Period____

Electricity and Gravity

1...Find the Force of Gravity between masses 9.00 g and 9.00 g at a distance of 3.00 m apart:

2...If the Distance between objects is quadrupled, by how much does the Force change?

3...If one mass is doubled by how much does the Force change?

4....If one mass triples and the other drops to a 1/4 by how much does the Force change?

5...If the distance is tripled and one mass is doubled by how much does the Force change?

6...If one mass is cut in half and the distance is doubled, by how much does the Force change?

7...What is the Electric Force between a -4.00C and a 2.00C particle a distance of 1.00 m apart?

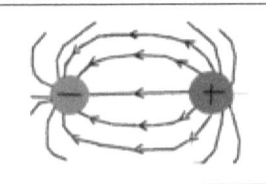

8...Is the Force Attractive or Repulsive?

9...What is the Distance between two 9.00C particles if their repulsion is 130. N?

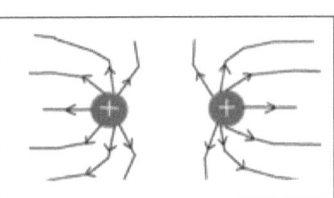

10...Find the Net Electric Force on the star labeled A.

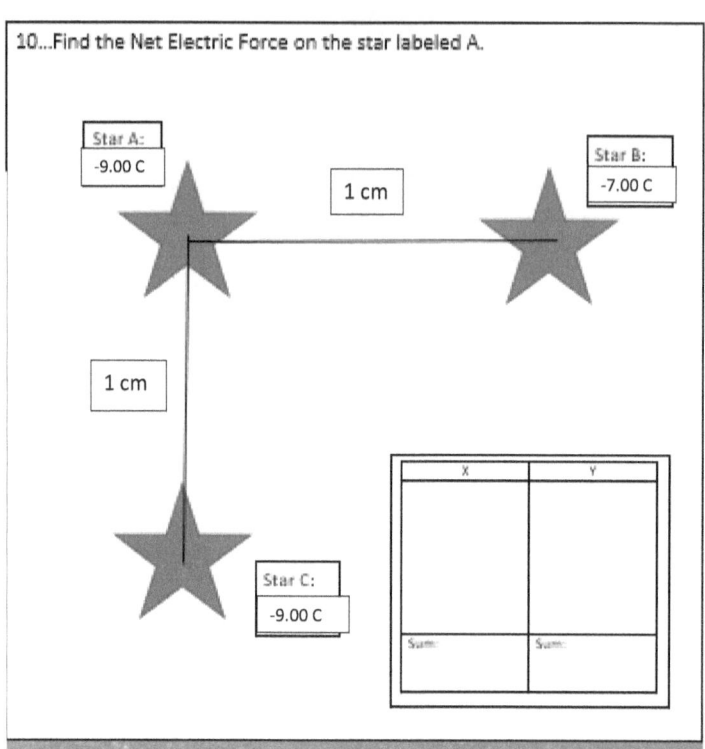

	X	Y
Sum:	Sum:	

Draw the Net Force triangle and find the magnitude and direction of the Force on star A:

Net Force:_____

11....What are the 4 Forces of nature?

    1:                              3:

    2:                              4:

12...Electric Field lines point_____ Positive Charge and _____ Negative Charge.

13...Draw Electric Field Lines around the particles:

Name_____ Period____

Electric Forces 4

1...Two metal spheres one with a Charge of $+5.2 \times 10^{-5} C$ and the other with a Charge of $+8.3 \times 10^{-5} C$ are separated by a Distance of 2.0 cm. Answer the following: **(In the following problems always use the result of the Force in Problem A)**

A...What is the Force between the two spheres?

B...Is the Force repulsive or attractive?

C...What is the new Force if the Distance between the two is cut to a fifth?

D...What is the new Force if the charge on both is doubled?

E...What is the new Force if the Charge on only one Charge is cut to a fifth?

F...What is the new Distance between the Particles if their new Force is eight times greater?

G...By how much must you increase the Distance between the two Particles for the Force to decrease three times?

2...A Proton and an Electron are separated by a distance of $3.3 \times 10^{-8} m$.

A...What is the Electric Force between them?

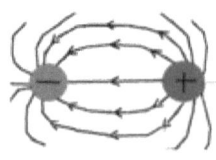

B...What is the Gravitational Force between them?

C…How much greater is the Electric Force compared with the Gravitational Force?

3…What is the Charge of two equal Particles if they are 5.00 m apart, and their Force is 6.00 N?

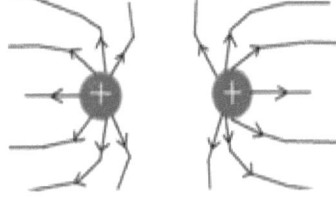

4...What is the Charge of two Particles if the Charge in one is six times greater than the other, both are negative and their separation is 3.00 m, and the Force between them is $7.8 \times 10^{-2} N$?

5...How many Electrons must be removed from a conducting sphere to give it a Charge of +7.0 C?

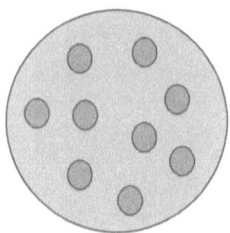

Name_____Period____

Electricity and Gravity

1...Find the Force of Gravity between masses 1.00 g and 9.70 g at a distance of 7.00 m apart:

2...If the Distance between objects is increased eight times, by how much does the Force change?

3...If one mass is tripled by how much does the Force change?

4....If one mass triples and the other drops to a 1/2 by how much does the Force change?

5...If the distance is tripled and one mass is quadrupled by how much does the Force change?

6...If one mass is cut to a third and the distance is tripled, by how much does the Force change?

7...What is the Electric Force between a 8.00C and a 1.00C particle a distance of 4.00 m apart?

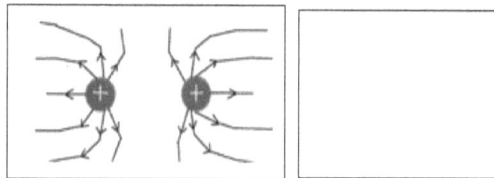

8...Is the Force Attractive or Repulsive?

9...What is the Distance between two 6.00C particles if their repulsion is 300. N?

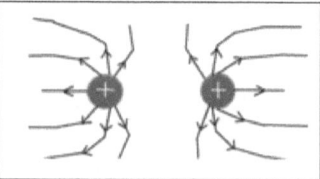

10... Find the Net Electric Force on the star labeled A.

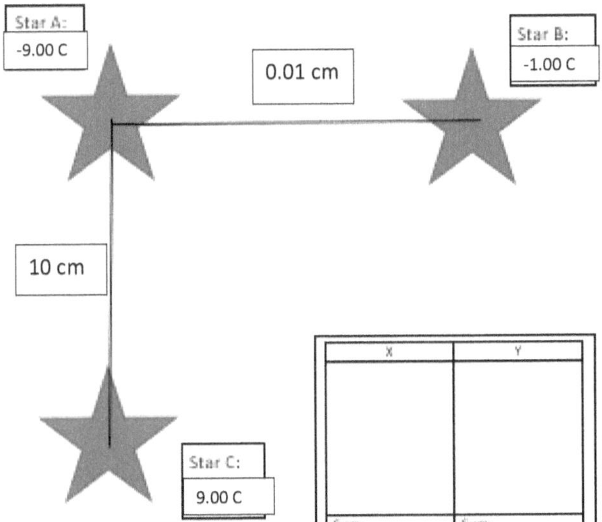

Star A: -9.00 C

Star B: -1.00 C

0.01 cm

10 cm

Star C: 9.00 C

X	Y
Sum:	Sum:

Draw the Net Force triangle and find the magnitude and direction of the Force on star A:

Net Force: _____

Name_____ Period____

Electric Forces Test

1...Two metal spheres one with a Charge of $2.2 \times 10^{-2} C$ and the other with a Charge of $2.3 \times 10^{-2} C$ are separated by a Distance of 2.0 m. Answer the following: **(In the following problems always use the result of the Force in Problem A)**

A...What is the Force between the two spheres?

B...Is the Force repulsive or attractive?

C...What is the new Force if the Distance between the two is doubled?

D…What is the new Force if the charge on both is dropped to a third?

E…What is the new Force if the Charge on only one Charge is cut to a quarter?

F…What is the new Distance between the Particles if their new Force is three times greater?

G...By how much must you increase the Distance between the two Particles for the Force to decrease five times?

2...A Proton and an Electron are separated by a distance of $1.3 \times 10^{-2} m$.

A...What is the Electric Force between them?

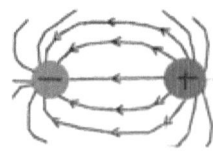

B...What is the Gravitational Force between them?

C...How much greater is the Electric Force compared with the Gravitational Force?

3...What is the Charge of two equal Particles if they are 3.00 mm apart, and their Force is 1.00 N?

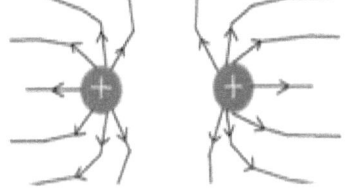

4...What is the Charge of two Particles if the Charge in one is two times greater than the other, both are negative and their separation is 3.00 mm, and the Force between them is 1.80 N?

Q =

2Q =

5...How many Electrons must be removed from a conducting sphere to give it a Charge of +2.0 C?

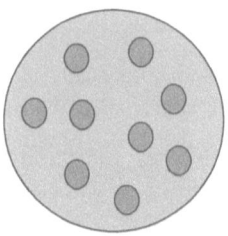

Name_____Period____

Electricity and Gravity

1...Find the Force of Gravity between masses 2.00 kg and 9.70kg at a distance of 2.00 m apart:

2...If the Distance between objects is increased nine times, by how much does the Force change?

3...If one mass is doubled by how much does the Force change?

4....If one mass doubles and the other drops to a 1/2 by how much does the Force change?

5...If the distance is tripled and one mass is doubled by how much does the Force change?

6...If one mass is cut to a third and the distance is quadrupled, by how much does the Force change?

7...What is the Electric Force between a 9.00C and a 1.00C particle a distance of 2.00 mm apart?

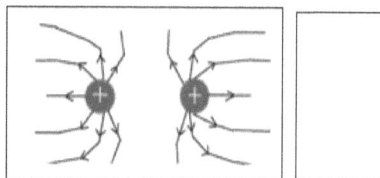

8...Is the Force Attractive or Repulsive?

9...What is the Distance between two 2.00C particles if their repulsion is 100. N?

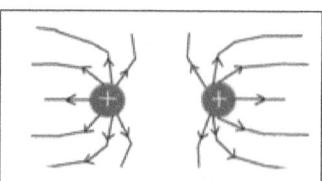

10...Find the Net Electric Force on the star labeled A.

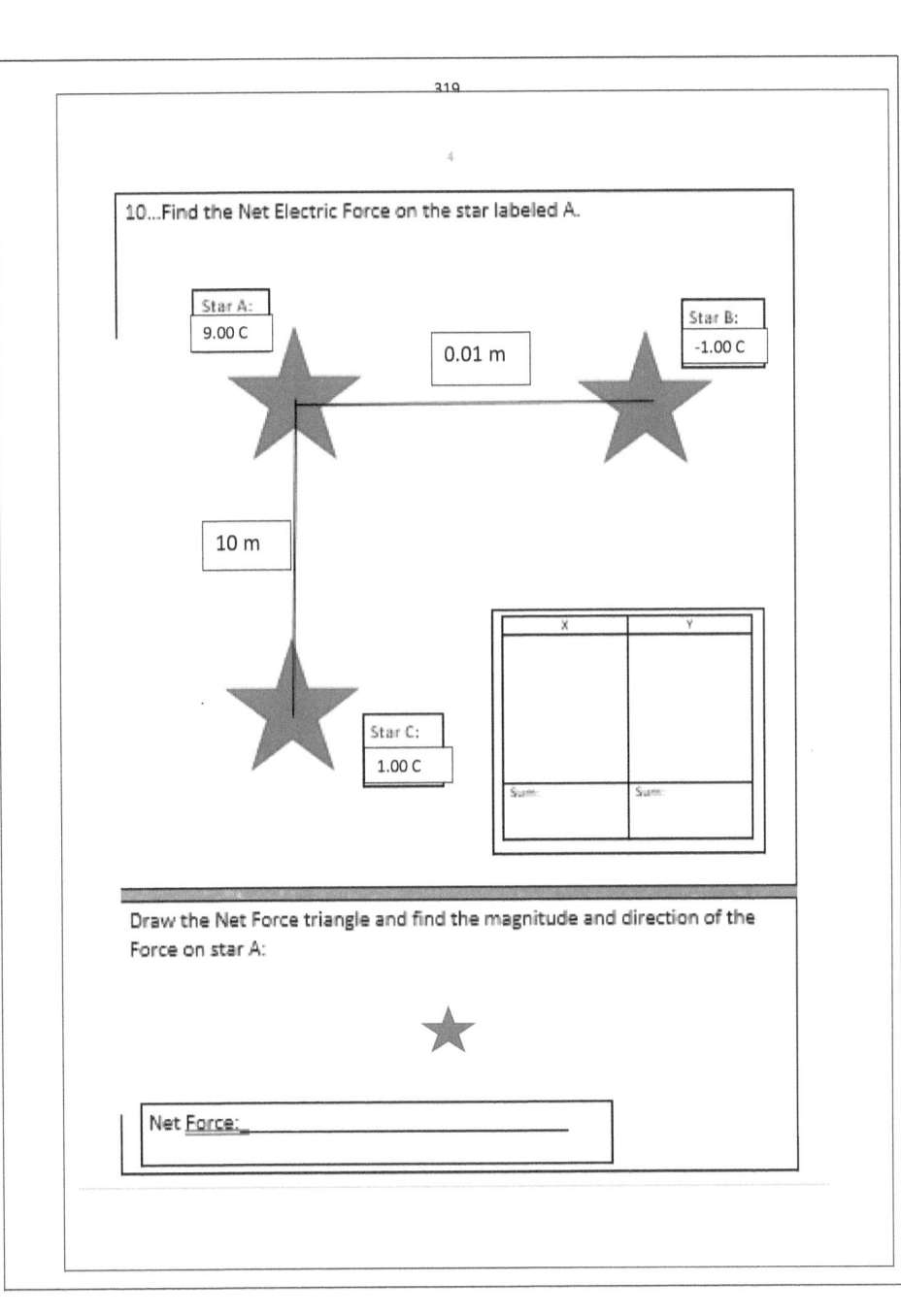

Draw the Net Force triangle and find the magnitude and direction of the Force on star A:

Net Force: _____

11….What are the 4 Forces of nature?

    1:                              3:

    2:                              4:

12…Electric Field lines point_____ Positive Charge and _____ Negative Charge.

13…Draw Electric Field Lines around the particles:

Name_____Period____

Vacuum Ray Tube

1…A heater of Ionized Gas shoots an Electron through a tube with Parallel Plates that have an Electric Field.

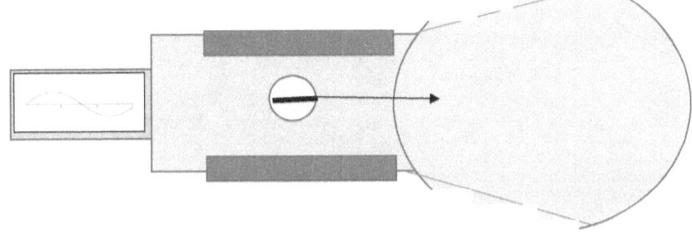

A…What must be the direction of Electric Field for the Electron to travel in a straight line?

B…What must be the Electric Field?

2...A heater of Ionized Gas shoots a Proton through a tube with Parallel Plates that have an Electric Field.

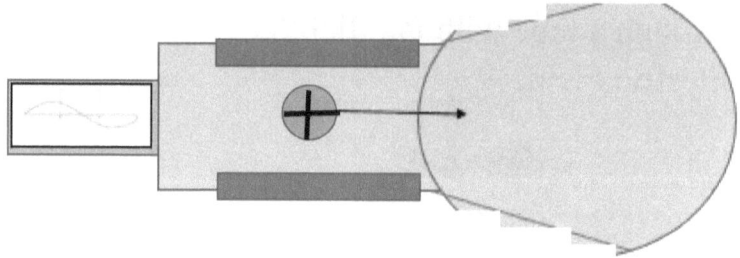

A...What must be the direction of Electric Field for the Proton to travel in a straight line?

B...What must be the Electric Field?

3...An Electron is fired at 1.00 m/s. It passes between Plates that are 0.01 m wide.

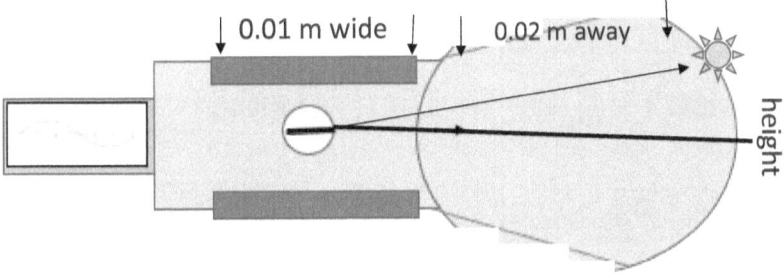

A...What is the Force on the Electron due to an Electric Field from the Plates of $1.14 \times 10^{-9} \left(\frac{N}{C}\right)$?

B...What is the Acceleration of the Electron in the y-direction inside the region between Plates?

Ignore Gravity

C…At what Height will the Electron hit the screen?

$(1/2)(a)(t1^2) + (a)(t1)(t2) =$ Distance

$(1/2)(\frac{F}{m})(\frac{W1}{Vx})^2 + (\frac{F}{m})(\frac{W1}{Vx})(\frac{W2}{Vx}) =$ Height

$(\frac{F}{m})(\frac{W1}{Vx})^2 (\frac{1}{2} + \frac{W2}{W1}) =$ Height

$a(\frac{W1}{Vx})^2 (\frac{1}{2} + \frac{W2}{W1}) =$ Height

4...An Electron is fired at 0.50 m/s. It passes between Plates that are 0.02 m wide.

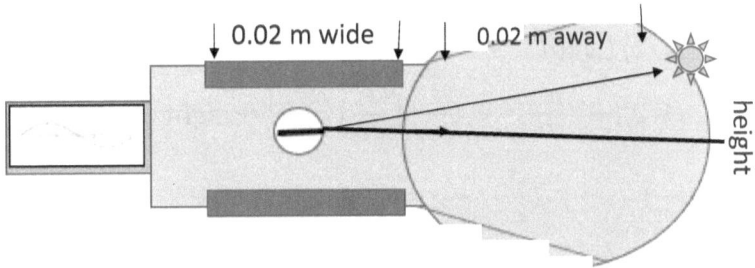

A...What is the Force on the Electron due to an Electric Field from the Plates of $5.70 \times 10^{-10} \left(\frac{N}{C}\right)$?

B...What is the Acceleration of the Electron in the y-direction inside the region between Plates?

Ignore Gravity

C…At what Height will the Electron hit the screen?

$(1/2)(a)(t1^2) + (a)(t1)(t2) = $ Distance

$(1/2)(\frac{F}{m})(\frac{W1}{Vx})^2 + (\frac{F}{m})(\frac{W1}{Vx})(\frac{W2}{Vx}) = $ Height

$(\frac{F}{m})(\frac{W1}{Vx})^2 (\frac{1}{2} + \frac{W2}{W1}) = $ Height

$a(\frac{W1}{Vx})^2 (\frac{1}{2} + \frac{W2}{W1}) = $ Height

5...Before the Electron is fired in a Vacuum Ray Tube it is accelerated over a Potential Difference of $1.0 \times 10^{-5} V$. After crossing the gap between the two charged plates the Electron goes through the center hole into the Ray Tube.

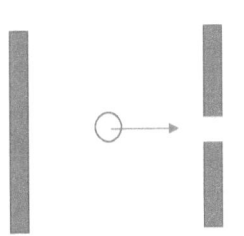

A...What is the Sign of the Charge of the Plate on the left?

B... What is the Sign of the Charge of the Plate on the right?

C...What is the gain in Kinetic Energy of the Electron?

D...What is the Final Velocity of the Electron when exiting the hole?

6...Before the Electron is fired in a Vacuum Ray Tube it is accelerated over a Potential Difference of $1.0 \times 10^{-4} V$. After crossing the gap between the two charged plates the Electron goes through the center hole into the Ray Tube.

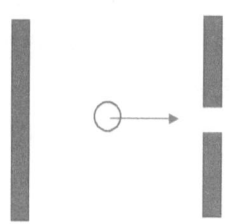

A...What is the Sign of the Charge of the Plate on the left?

B... What is the Sign of the Charge of the Plate on the right?

C...What is the gain in Kinetic Energy of the Electron?

D...What is the Final Velocity of the Electron when exiting the hole?

Name_____ Period_____

Circuits:

In the year 1752 Benjamin Franklin discovered Electricity from the flow of Electrons from a cloud to his kite through a lightning strike. Benjamin Franklin was also one of the Founding Fathers of the United States.

Electricity is the flow of Electrons through surfaces in which these Particles are free to move such as Conductors. Metals are good Conductors while rubber, for example is an insulator.

In the late 1799 the Italian Inventor called Alessandro Volta built the world's first battery. The word Voltage is in honor of him.

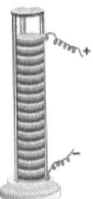

He had the idea of making a random tube with discs of two metals such as Zinc and Copper separated from each other by cardboard soaked in brine. The brine helps speed up the reaction of electrons moving from Zinc to Copper.

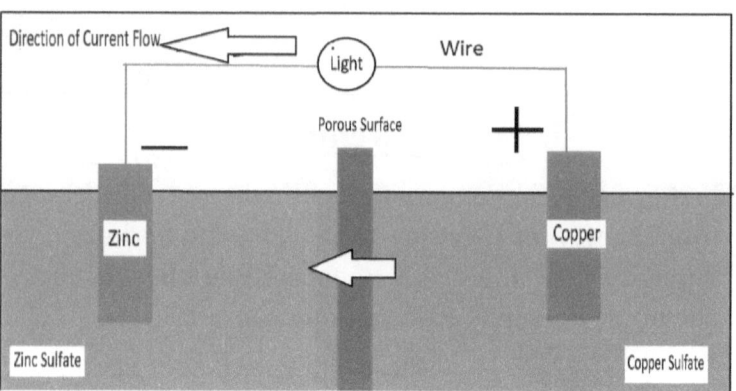

In the figure above there is a picture of how the metals look after the reaction has taken place for a while.

In the figure on the left there is a similar battery but not in a tube but rather using two separate containers. When a bar of Zinc is placed in a Zinc Sulfate Solution, the Zinc Atoms react with the sulfate leaving the metal. In the other container with a bar of copper inside of a Copper Sulfate Solution, the Copper atoms in the solution move towards the surface of the metal away from the sulfate. This process generates a flow of electrons from the negative zinc side to the positive copper side.

In the 1780s The French Physicist Charles Augustine Coulomb investigating electricity discovered that:

Voltage = (Current)(Resistance)

The unit for Charge is the Coulomb in honor of him. Current is a measure of Coulombs per second while Voltage is Energy per Charge.

Resistance is a measure of how much a material resists the flow of Electric Current and is measured in Ohms.

Volts = J/C

Current = C/s

Resistance = Ω

Resistance = $\frac{\rho L}{A}$

Where:

ρ = Resistivity

L = Length

A = Cross Sectional Area

Each material has a Resistivity. Using the equation above, its Resistance can be found. The longer the material the greater the Resistance since Electrons need to move farther. The thicker the material the lower the Resistance since Electrons have a larger area to move into the material.

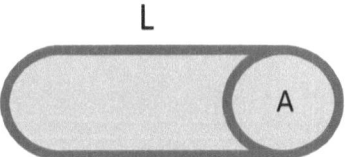

Solve for the resistance of the following materials with Length 10.0 m and cross sectional area 2.00 m^2:

Material	Resistance
Cu	
Fe	
Ag	
Al	
Ni	
Cu-Fe	
Cu-Ag	
Al-Ni	

Fill in the blank with the words: greater or smaller:

The _____ the cross sectional area of a material the _____ is its resistance. The _____ is the length of the material the _____ is its resistance.

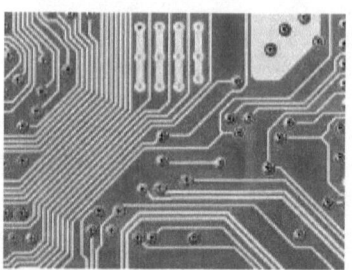

There are two types of circuit: Series and Parallel. Circuits can also be a mix of both.

These are the equations for calculating Voltage, Current, and Resistance of the parts of a Series Circuit:

Voltages add

Currents the same

Resistances add

Circuits everywhere in my computer

These are the equations for calculating Voltage, Current, and Resistance of the parts of a Parallel Circuit:

Voltages the same

Currents add

Total Resistance = $(1)/((1/R1) + (1/R2) + \ldots)$

Thanks to Physics I have a Cell Phone!

1- Fill out the Table below:

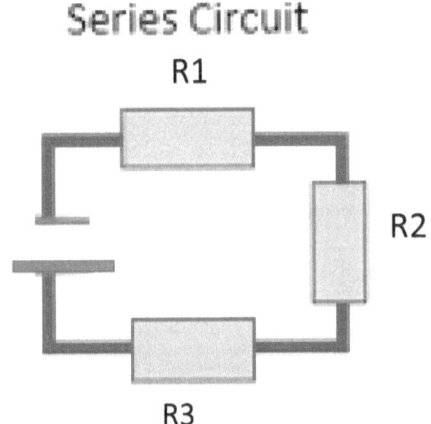

Resistor	Voltage	Current	Resistance	Power
1			11.0 ohms	
2			13.0 ohms	
3			11.5 ohms	
Total	70.0 V			

2-Second Table

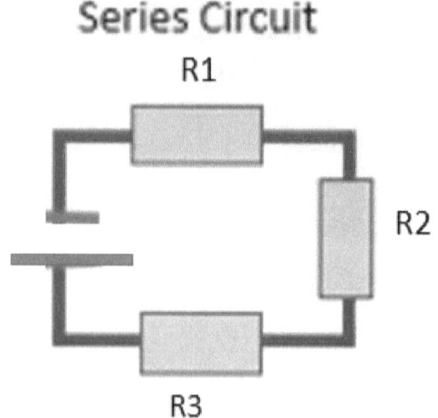

Resistor	Voltage	Current	Resistance	Power
1			1.0 ohms	
2			1.0 ohms	
3			1.5 ohms	
Total		80.0 A		

3-Third Table

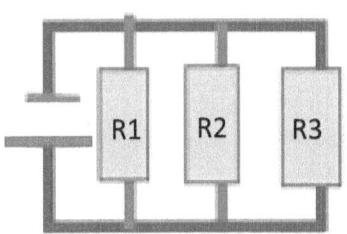

Parallel Circuit

Resistor	Voltage	Current	Resistance	Power
1			10.0 ohms	
2			9.0 ohms	
3			12.5 ohms	
Total	5.0 V			

4-Fourth Table

Parallel Circuit

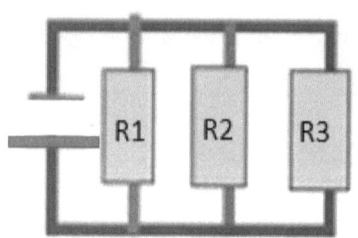

Resistor	Voltage	Current	Resistance	Power
1		45.0 A	10.0 ohms	
2			13.0 ohms	
3			14.5 ohms	
Total				

Name_____Period_____

Electricity and Magnetism

In the beginning it was the Big Bang from which the universe was formed:

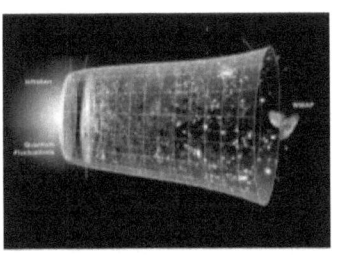

After the Big Bang the grand cosmic unified force separated in to the four forces present in the universe today. These forces are:

1) **Gravitational**: Causing masses to attract other masses.
2) **Electromagnetic**: Giving charge to particles.
3) **Weak**: Breaks the nucleus of atoms.
4) **Strong**: Keeps the nucleus of atoms together.

Everything began at Big Bang. All things have a single common origin!

The image on the right illustrate the Fundamental particles of the universe. The four forces of nature are mediated by Bosons:

The Gravitational Force needs the Graviton. The Electromagnetic Force needs the Photon. The Weak Force needs the Z and W Bosons. The Strong Force needs the Gluons.

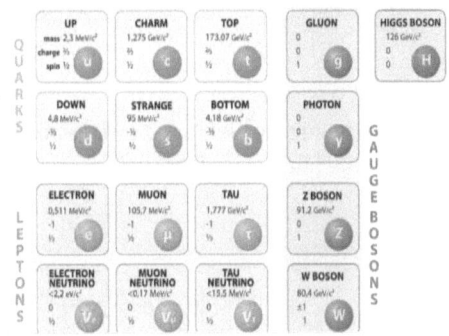

The Gravitational Force between two masses can be found using the following equation:

$$F = \frac{G\,(M1)(M2)}{r^2}$$

Where G is the Gravitational Constant equal to 6.67x10^-11 N (m^3)/(kg)(s^2)

M1 and M2 are the masses 1 and 2 while r is the distance between their centers.

1...Find the Force of Gravity between masses 3.00 kg and 6.00 kg at a distance of 1.00 cm apart:

5...If the distance is tripled by how much does the Force decrease?

6...If one mass is cut in half and the distance is double, by how much does the Force change?

The inverse square law present in the Gravitational and Electric Force obeys the following:

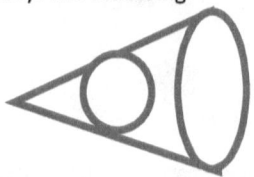

Isaac Newton discovered the relation with the fact that the surface area of a sphere is $4\pi r^2$. The distance away from the center of an object's gravity is proportional to 1/(r^2).

2...If the distance between objects is cut in half, by how much does the Force increase?

3...If the one mass is doubled by how much does the Force increase?

4....If both masses tripled by how much does the Force increase?

Electric Field Lines are a field of photons emanating from the particles. The Field lines point towards the Negative Charges and away from the Positive charges.

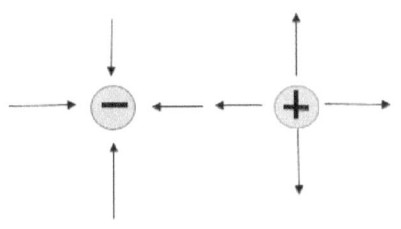

These are the Electric Field Lines between two opposite charged Particles. The Force between them is attractive:

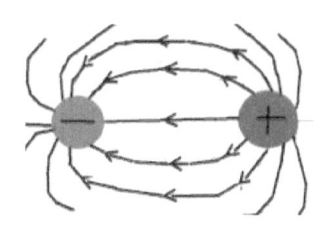

7...What is the Electric Force between a 3.00μC and a -9.00μC Particles a Distance of 4.00 mm apart?

8...Is the Force attractive or Repulsive?

These are the Electric Field Lines between two like charged Particles. The Force between them is repulsive:

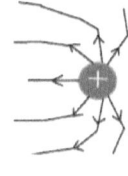 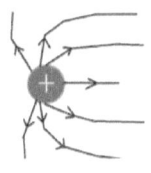

$F = kQq/r^2$

$k = 9 \times 10^9 \, Nm^2/C^2$

Q and q being the two Charges

r being the Distance between them

The Electric Force between charges is the following:

9...What is Distance between two 4.00C particles if their repulsion is 10.0 N?

10...Find the Net Electric Force on the star labeled A.

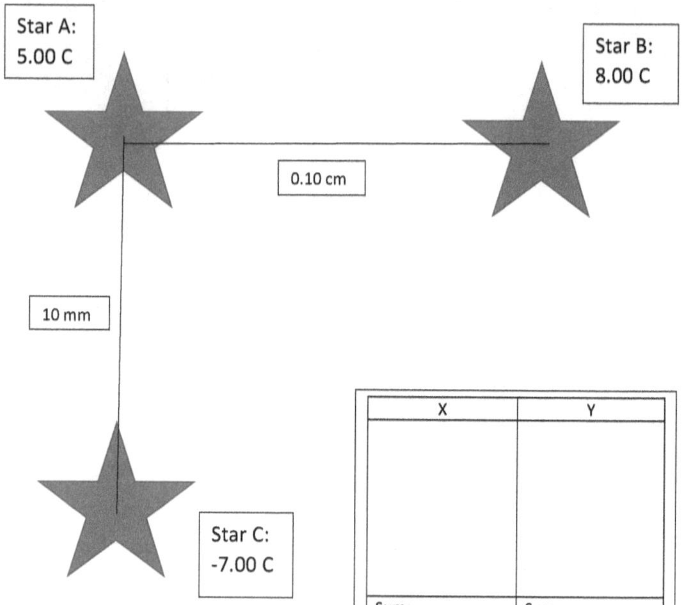

Star A: 5.00 C

Star B: 8.00 C

0.10 cm

10 mm

Star C: -7.00 C

	X	Y
Sum:		

Draw the Net Force triangle and find the magnitude and direction of the Force on star A:

Net Force:_____

Light is an Electromagnetic Wave: The Electric and Magnetic Fields are always 90 degrees from each other.

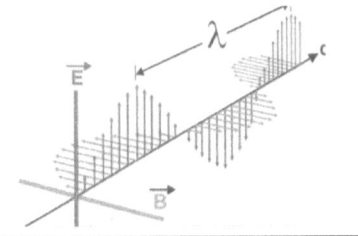

11...Label the colors of light from most energetic to least energetic:

The Electromagnetic Spectrum is composed of several forms of light:

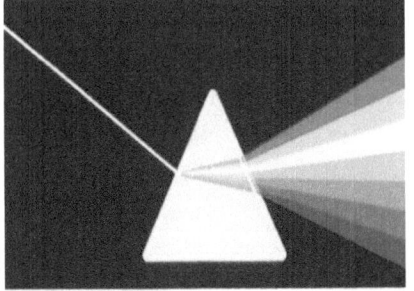

Isaac Newton discovered that color white is composed of all the colors of the rainbow. These colors can be broken when white light is passed through a prism.

Electric Field between plates:

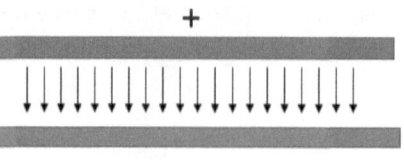

12....What is the Electric Field coming from an 8.00C Particle at a Distance of 3.00 m away?

16....What is the Electric Field coming from a -6.00C Particle at a distance of 2.00 m away?

13...Does the Field point towards or away from the Particle?_____.

17...Does the Field point towards or away from the Particle?_____.

14...If an 8.00C Particle is placed at that Distance from the 8.00 C Particle what will be the Force that it will experience?

18...If a -6.00C particle is placed at that distance from the -6.00 C particle what will be the Force that it will experience?

15....Will that Force be attractive of repulsive?_____.

19....Will that Force be attractive of repulsive?_____.

Electricity and Magnetism

Lonyfaryondy: Today were going to talk about electricity and some electronics. There are three components of a circuit essential for our experience here today. Besides the Battery and the wire they are the Resistor, Capacitor, and Inductor.

Current through the wires of a circuit are caused by an Electric Potential Difference (Voltage), which are an unbalance of charges that leads to their flow. When inserting a copper metal and a zinc metal in a potato, connected in series with other potatoes or with a device, electricity is allowed to flow. The positive zinc ions from the zinc metal moves out of the metal into the potato leaving two electrons that flow through the wire towards the copper metal which attracts other positive ions in the potato. The copper metal becomes positively charged and the zinc negatively charge. This process leads to a Battery and a flow of charges which can be used to turn on devices such as a light bulb.

Potato Battery

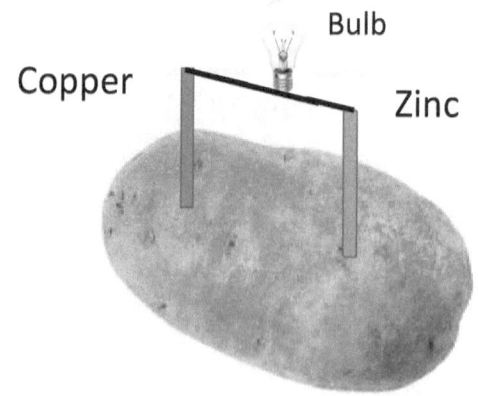

Current

Currents are flow of charges. The conventional current is understood to be the flow of positive holes which are regions in matter where Electrons are missing. In reality current is the flow of Electrons but in the 1800s it was believed that current was the flow of positive charges, and when scientists realized that it was the negatively charged Electrons that flowed they decided to keep the idea that Current is the flow of positive holes instead which is thought to be regions of missing Electrons. Conventional positive Current flows in the opposite direction of the Electron flow.

Resistor

Resistors are the parts of the Circuit which resist the flow of Current which are used to control the amount of flow of the charges per time. The wider and shorter the resistor the more charges can flow through it, while the narrower and longer the resistor the less charges can flow through it at a given time.

Capacitor

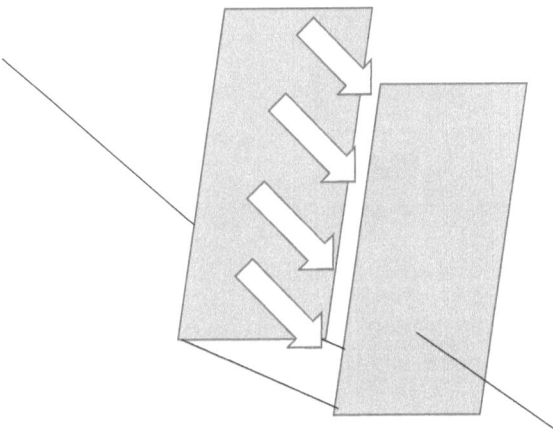

Yellow arrows show direction of Electric Field between the plates

A Capacitor is used to store a given amount of Energy and Potential Difference on each plate of opposite charge as the other. When needed, the stored Energy in the Capacitor is released to be used for a short time. The Capacitor can then be recharged for later usage.

It is possible to increase the Capacitance of a Capacitor by inserting a Dielectric Material allowing a greater amount of charge to be stored per Voltage. The Dielectric Material reduces the Electric Field between the Plates and with a smaller Voltage there is an increase in Capacitance. The equation of Capacitance is Charge/Voltage. With a Dielectric Material there is less Voltage which means greater Capacitance.

Inductor

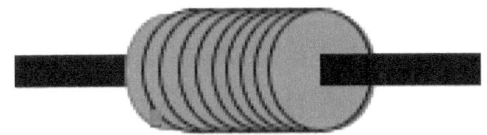

The Inductor is used to reverse Current or to generate an Alternating Current.

When inserting a magnet into an inductor a Current is generated according to the figure below.

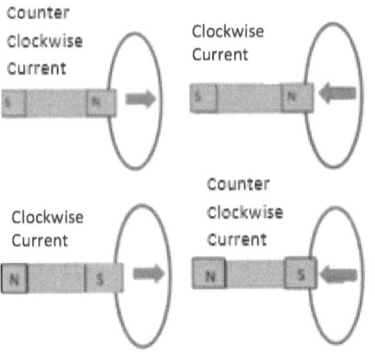

The induced Current in the Inductor is always an act of nature that opposes the change in Magnetic Flux in its inside. The Current inside an Inductor also generates a Magnetic Field that opposes the Magnetic Field of the Magnet that is being inserted or removed from the entrance of the coil.

Diode

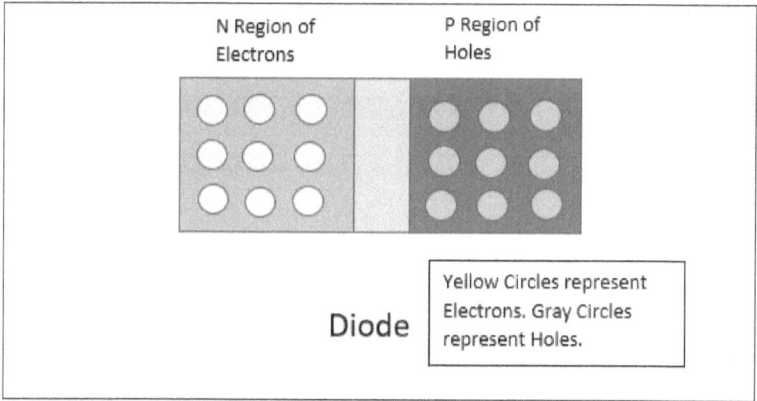

A Diode is a device that only allows current in one direction. The N Region of the Diode is made of a semiconductor material with grains of Atoms of another material with an extra Electron in its Valence Orbit. These extra Electrons from these atoms gives the N region of Diode a negative charge. The P Region of the Diode is made of the same semiconductor material but this time with grains of Atoms with one less Electron in the Valence Orbit. These extra Positive Charged Holes give the P Region a positive charge.

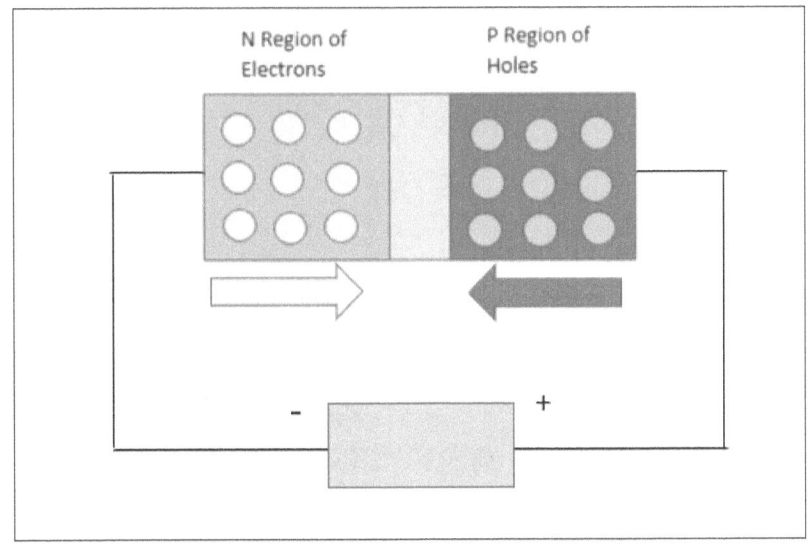

When a Battery is wired like in the previous figure the positive side of the Battery repels the Positive Holes in the Diode and Attract the Electrons which allows Current to Flow.

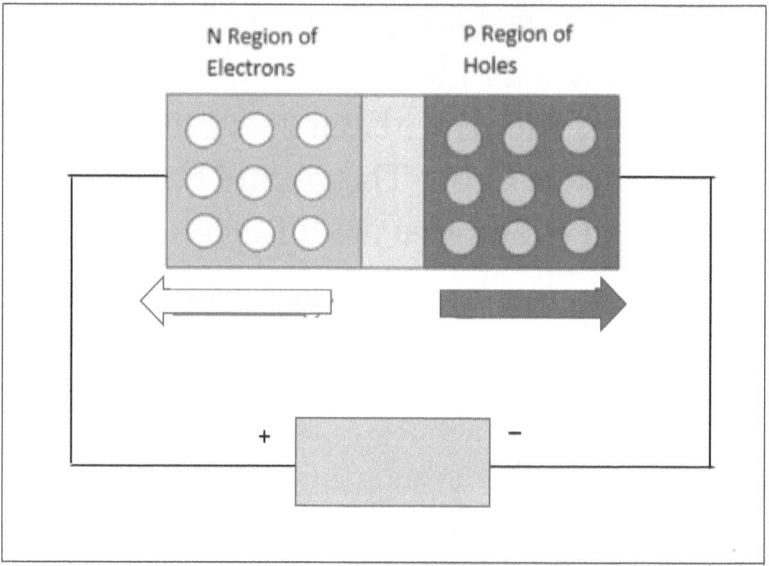

When a Battery is wired like in the figure above the Negative side of the Battery attracts the Positive Holes, and the Positive side of the Battery attracts the Electrons and this makes the charges in the Diode in being pulled in two different direction preventing the flow of Current.

Transistor

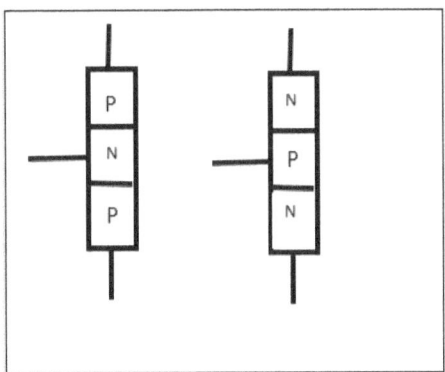

Transistors are made of two Diodes glued together. The combination inside of Transistors can be regions of npn or pnp as shown:

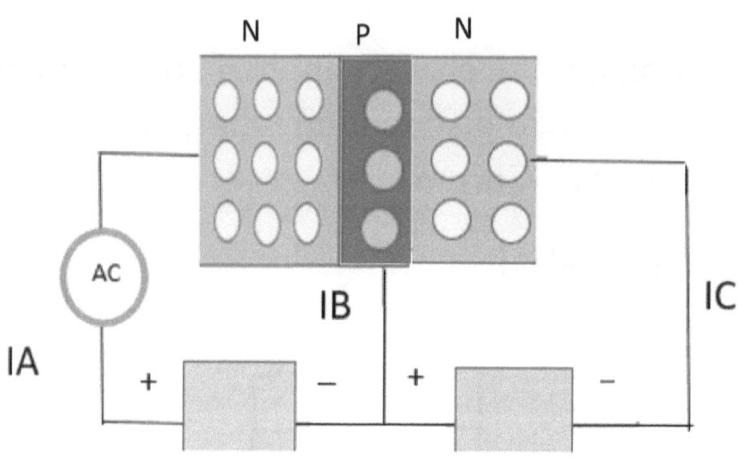

An Alternating Input Current IA pulls and pushes on the Electrons in the N Region. When Electrons are pulled the Positive Holes in the P Region are forced out into Current IB, which in effect increases the amount of Current in IC increasing the flow of charges which amplifies the Positive Signal at Current IC. When Electrons are Pushed in the N Region the Positive Holes from IB go into the P Region amplifying the loss of Current in IC. The result is an

Amplified Negative Signal in IC with a higher Amplitude.

Input Signal

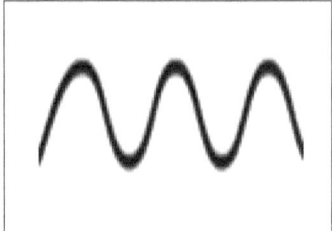

Output Signal
Higher Amplitude

Transistors can also be used as on and off switches highly useful in computer binary language of 0s and 1s, on and off signals. When there is an increase in Current it is an on signal, and when there is a decrease in Current there is an off signal. There is over a billion Transistors in the Microprocessor of a Cell Phone.

Alternating Current

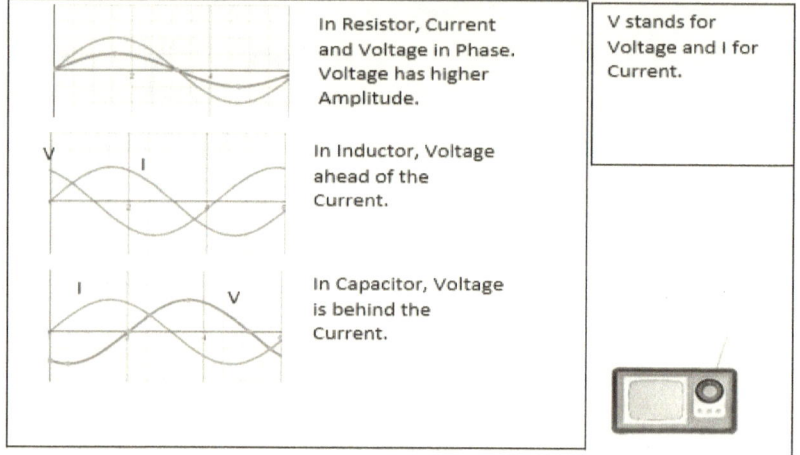

The figure above shows that a combination of an Inductor, Capacitor, and Resistor gives rise to an alternating Current in a Circuit. If an Antenna is placed in the Circuit and the Circuit is also grounded, the device becomes a Radio. The Frequency of the Alternating Current can then be changed by changing the Capacitance of the Capacitor. The

Capacitance of the Capacitor is changed when we try to move through the Radio Stations captured by the Radio. If the Frequency matches the Frequency from a Radio Station, transmission is made and you are able to hear what is being sent in the air by the Antenna of the Radio Station.

The world is surrounded by a sea of Radio Waves from all the telecommunications in the modern civilization.

Speaker

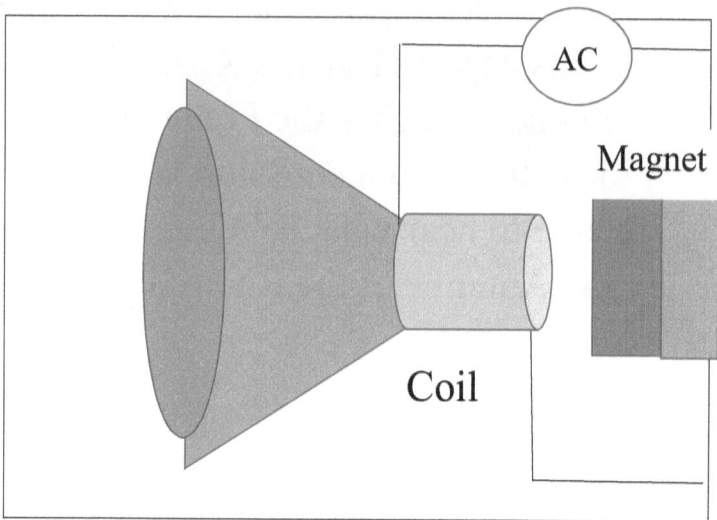

A Speaker converts an Alternating Current into Sound by forcing the Coil up and down against and towards the Magnet. This up and down motion vibrates a membrane that vibrates the particles in air leading to Sound.

Microphone

The Microphone works in reverse. Sound forces a membrane up and down, which moves the Coil above a Magnetic Field up and down, leading to an Alternating Current converting Sound into Electricity.

Cathode Ray Television

A Cathode Ray Television works using the principle shown below:

Electrons are fired from the heating filament at point C and accelerated in a vacuum tube through a Potential Difference.

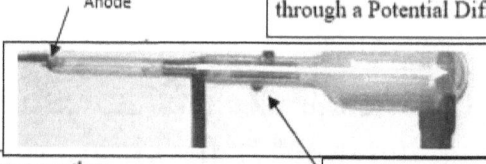

Anode

The Electrons are then deflected by Magnets or by an Electric Field to hit a specific point on the screen.

Deflecting Plates

The amount of deflection of the Electrons can be adjusted by changing the field for Electrons to hit different parts of the screen.

Anode

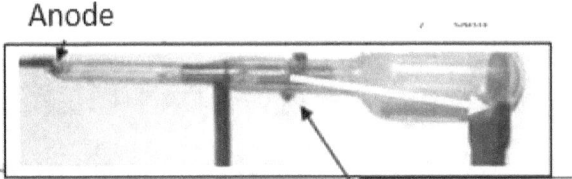

The Electrons are fired at the Phosphor Screen. Each Pixel on the screen contains a Red, Blue, and Green portion. When the Electrons hit the Green a Green dot is seen in the screen, and the same for Red and Blue. A combination of all the Pixels and the colors generate an image on the Screen.

In order to generate a complete picture, the Electrons are deflected scanning the screen from right to left and then moving downwards. The scan is done many times a second leading to an image that moves.

Flat Screen TVS use LED lights that turn on or off in each Pixel in the screen to generate the image which is made of a combination of these Pixels and colors.

Electromagnetic Fields and Forces

An Introduction to Maxwell's Equations:

Let us think of a simple magnet that is a dipole never a monopole. It has a N, North Pole, and S, South Pole. The magnetic field points away from the North and towards the South Pole.

Inserting a magnet with its N side into a loop will cause a current counterclockwise.

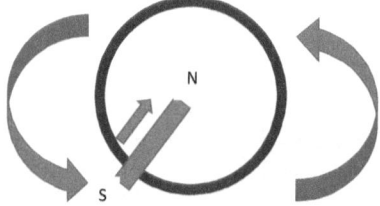

Using the same magnet and moving away from the loop will cause a current clockwise.

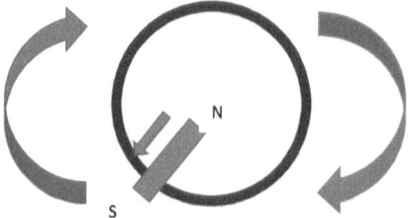

That is because nature abhors change.

In both cases there is a change in magnetic flux. The current on the wire tries to preserve the magnetic field that was before.

In a loop, a current clockwise will cause a magnetic field into the page.

In the same loop, a current counterclockwise will cause a magnetic field out of the page.

Inserting a magnet inside a loop with its S side will cause a current clockwise.

Draw what is going on according to the statement above.

Moving the same magnet away from the loop will cause a counterclockwise current.

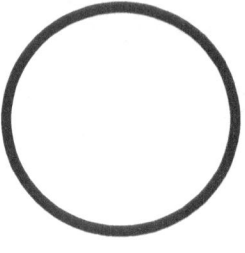

Draw what is going on according to the statement above.

A change in magnetic field causes an electric field.

Similarly, a change in an electric field will cause a magnetic field.

An electric field diverges from a source.

A magnetic field curls around itself, and its divergence is zero.

There are no magnetic monopoles. A magnet will always have a North side followed by a South side.

An electric field can be monopoles such as a single negative particle.

An Introduction to Maxwell's Equations:

Maxwell's Equations are able to explain all of Electrodynamics. I start this book with a quick summary.

Let us think of a simple magnet that is a dipole never a monopole. It has a N, North Pole, and S, South Pole. The magnetic field points away from the North and towards the South Pole.

Inserting a magnet with its N side into a loop will cause a current counterclockwise.

Using the same magnet and moving away from the loop will cause a current clockwise.

That is because nature abhors change.

In both cases there is a change in magnetic flux. The current on the wire tries to preserve the magnetic field that was before.

In a loop, a current clockwise will cause a magnetic field into the page.

In the same loop, a current counterclockwise will cause a magnetic field out of the page.

Inserting a magnet inside a loop with its S side will cause a current clockwise.

Moving the same magnet away from the loop will cause a counterclockwise current.

A change in magnetic field causes an electric field.

Similarly, a change in an electric field will cause a magnetic field.

An electric field diverges from a source.

A magnetic field curls around itself, and its divergence is zero.

There are no magnetic monopoles. A magnet will always have a North side followed by a South side.

An electric field can be monopoles such as a single negative particle.

When a positive particle moves in a region of magnetic field it experiences a force as shown in the right-hand rule:

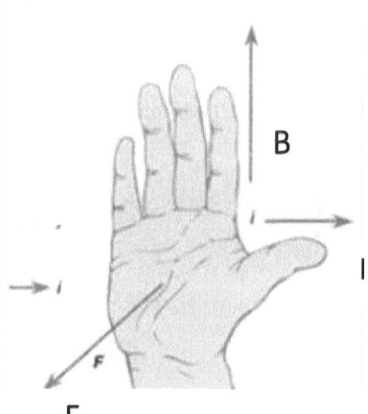

If the particle is negative the direction of the Force is opposite of what is shown:

Four fingers in the direction of Magnetic Field, Palm in the direction of Force, and Thumb in the direction of Current.

The two ways that the right-hand rule can be used is with the thumb pointing in the direction of the current I, the four fingers in the direction of the Magnetic Field, and the palm in the direction of the Force which applies for the flow of positive charges. For negative charges the Force will point in the opposite direction.

Black arrows in the direction of current in a wire

Green left and right arrows in the direction of Force

From a magnet the Magnetic Field points away from the North Pole and towards the South Pole.

Another use of the right hand rule is for coils where the four fingers point in the direction of the current, and the thumb in the direction of the Magnetic Field. A dot is used for a Magnetic Field out of the page while an x is used for a Magnetic Field into the page.

When a current runs through a coil it generates a magnetic field as shown:

That is using the right-hand rule once again

Thumb in the direction of Magnetic Field and Four Fingers in the direction of Current in Coil.

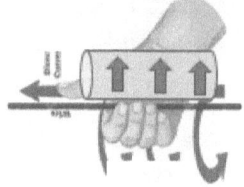

Magnetic Field from a Solenoid.

An Electric Motor uses current through a wired loop in a region of a Magnetic Field to cause a Force on both sides of the loop with current in two different directions leading to a rotation. To allow the rotation to continue beyond 90 degrees the source of current reverses the direction of the current causing the rotation to continue as long as the Alternating Current continues to flow in the loop.

A wire with current in the direction I in a region with a Magnetic Field will experience a Force F. This is the basis of how Electric Motors work as shown.

That is how an electric motor works:

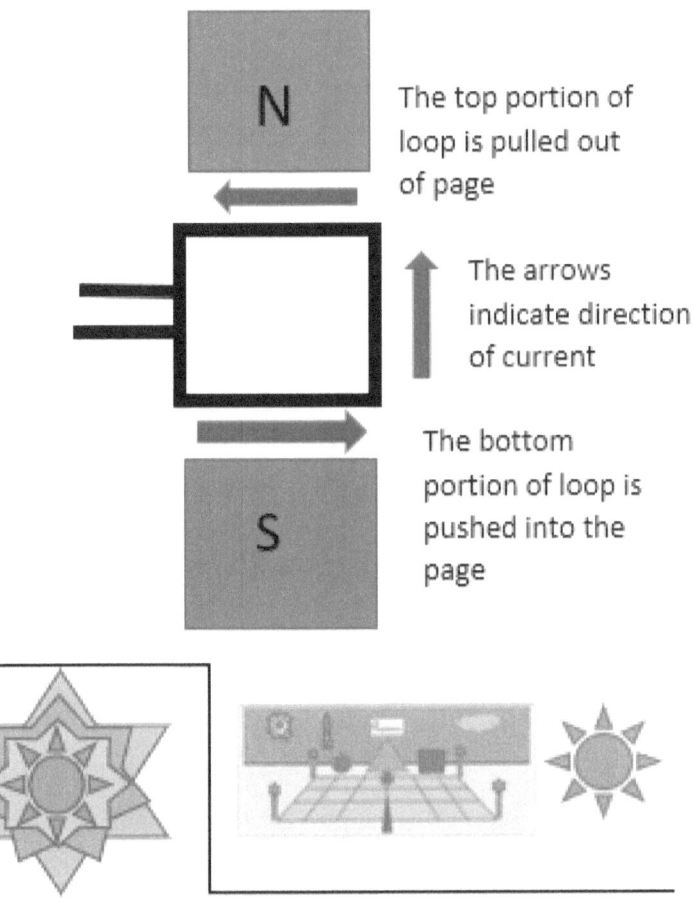

The top portion of loop is pulled out of page

The arrows indicate direction of current

The bottom portion of loop is pushed into the page

Induction Practice

Find the induced current in the following examples:

1...Coil flat on a region of magnetic field out of the page. The field decreases: <u>**CURRENT COUNTERCLOCKWISE**</u>

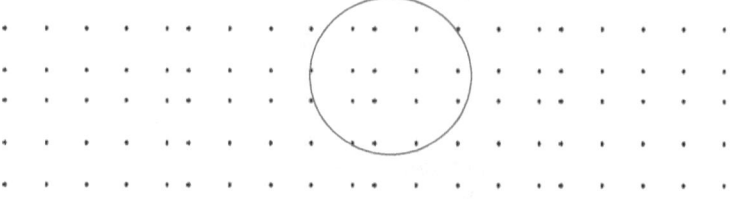

2.... Coil flat on a region of magnetic field out of the page. The field increases: <u>**CURRENT CLOCKWISE**</u>

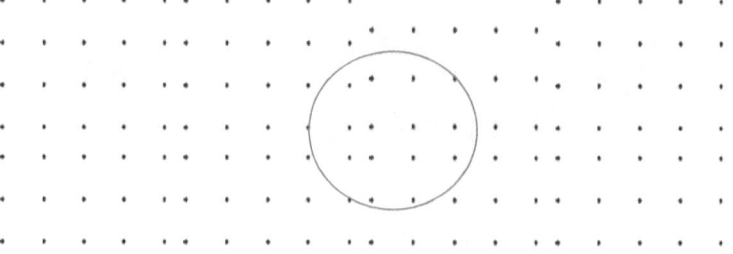

3… Coil flat on a region of magnetic field into the page. The field decreases: **CURRENT CLOCKWISE**

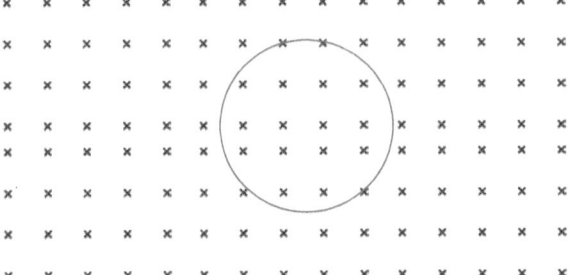

4… Coil flat on a region of magnetic field into the page. The field increases: **CURRENT COUNTERCLOCKWISE**

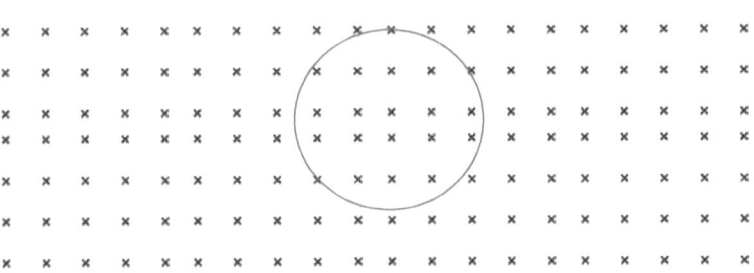

5…Coil travelling from the left into a magnetic field out of the page: **CURRENT CLOCKWISE**

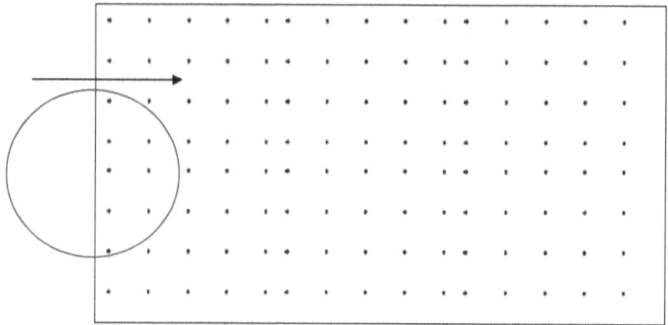

6…Coil travelling to the right leaving the magnetic field out of the page: **CURRENT COUNTERCLOCKWISE**

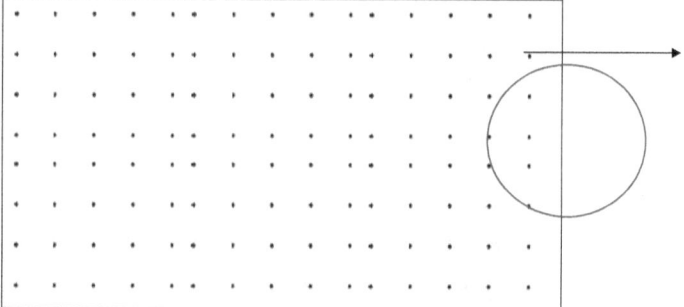

7… Coil travelling from the left into a magnetic field into the page: **CURRENT COUNTERCLOCKWISE**

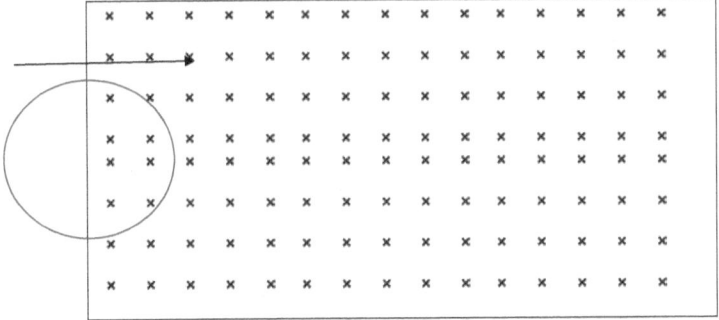

8… Coil travelling to the right leaving the magnetic field into the page: **CURRENT CLOCKWISE**

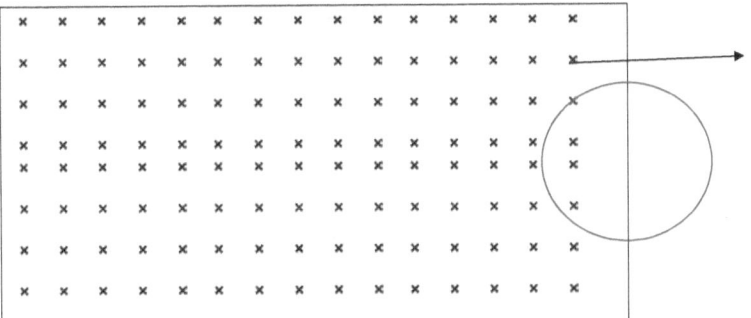

9...Coil travelling inside a region of magnetic field to the right: **NO CURRENT**

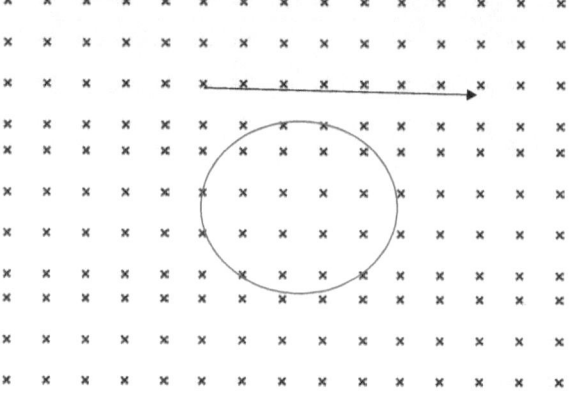

10... Coil travelling inside a region of magnetic field to the left: **NO CURRENT**

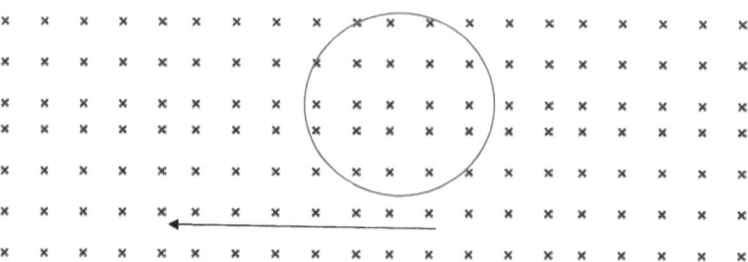

11… Coil travelling inside a region of magnetic field to the right: **NO CURRENT**

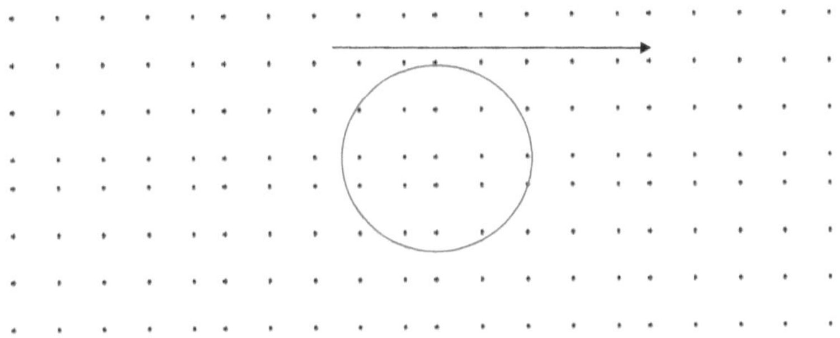

12… Coil travelling inside a region of magnetic field to the left: **NO CURRENT**

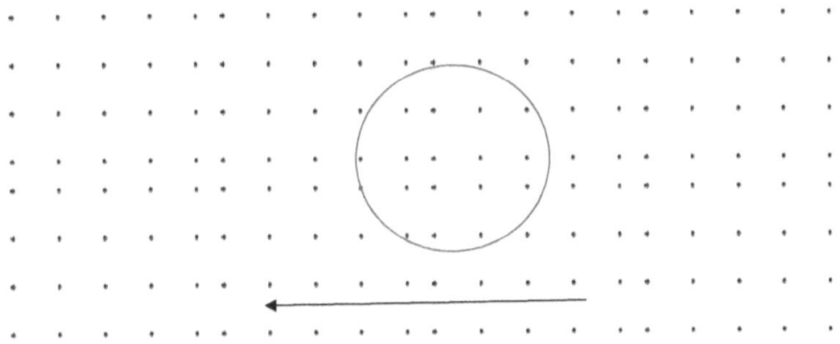

What is the induced current on the other coil?

A…

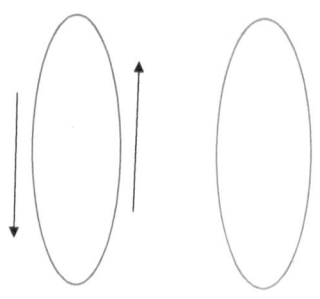

With a pulse of current in one loop there is a generation of an opposite current induced in the other loop.

B…

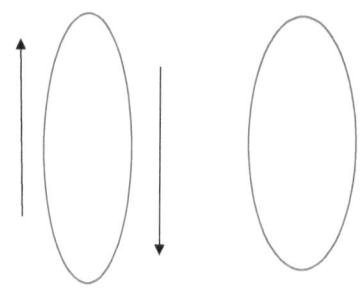

Continuous induction can only occur through pulses or changes in the current in one loop that affects the other at that given instant.

C…

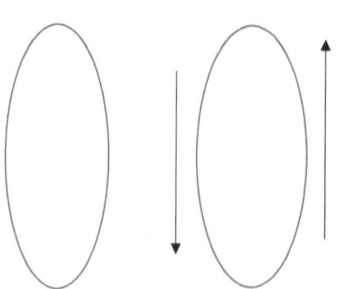

D…

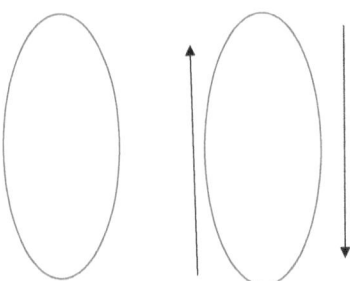

If the pulse of current in one coil is clockwise the induced current on the other loop will be counterclockwise and if the pulse is counterclockwise the induced current on the other loop is clockwise.

Vloudel then opened another book and read about forces due to current in a Magnetic Field.

When a wire carries a current a Magnetic Field is generated that curls around the wire like in the figure below using the right hand rule.

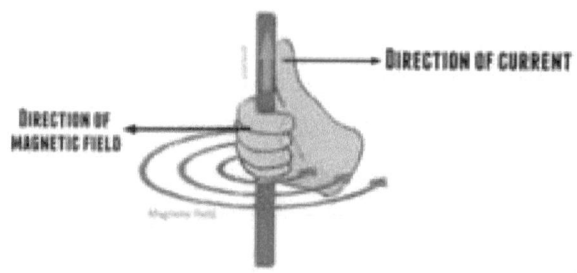

This phenomenon is explained by the Maxwell Equation $\nabla \times B$ which states that a change in Electric Field generates a Magnetic Field. When a current is flowing through the wire a Magnetic Field curls around the wire.

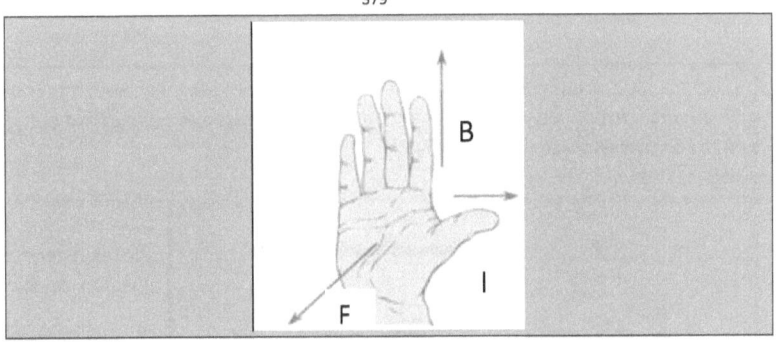

When a current flows through a wire that is present in a region of Magnetic Field, the direction of the force can be calculated by using Fleming's Left Hand Rule or the Right Hand Rule as show above.

Vloudel then showed Zeno some of the worksheets in Ogo's Book with answers:

Label the direction of the forces on the wires with current flowing in the direction of the arrows exposed to the magnetic field as shown:	An x shows Magnetic Field into the page and a dot shows a Magnetic Field out of the page.

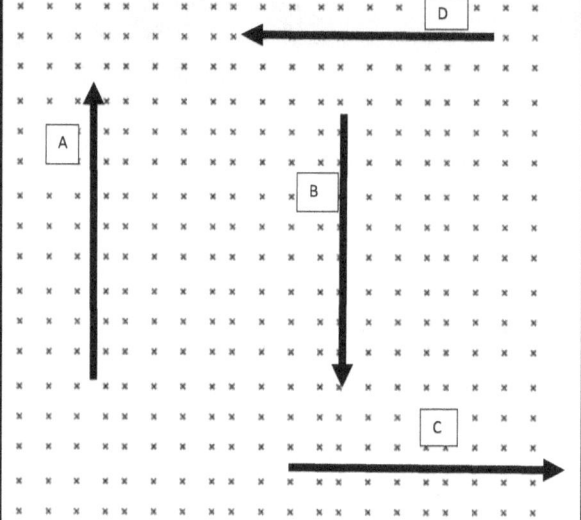

Label the force that the wire experiences as Left, right, up, or down:

A:__Left_____

B:__Right_____

C:__Up_____

D:__Down_____

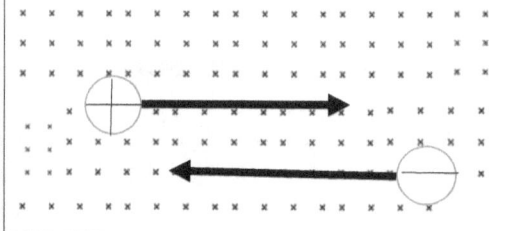

Positive Charge:____Up_____

Negative Charge:____Up_____

The results of the right hand rule are flipped for negative charges.

| Label the direction of the forces on the wires with current flowing in the direction of the arrows exposed to the magnetic field as shown: | An x shows Magnetic Field into the page and a dot shows a Magnetic Field out of the page |

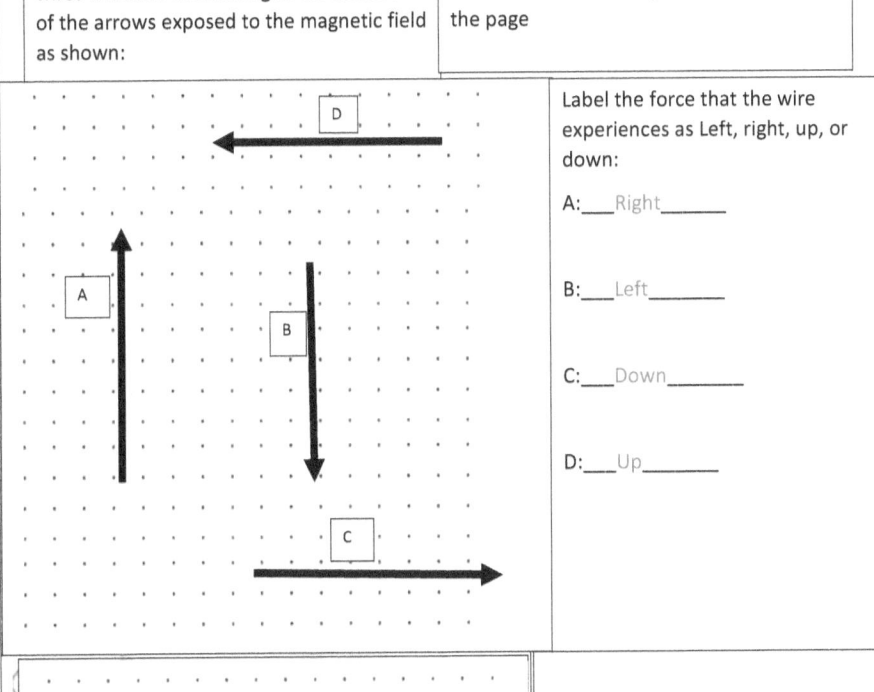

Label the force that the wire experiences as Left, right, up, or down:

A:___Right_____

B:___Left_____

C:___Down_____

D:___Up_____

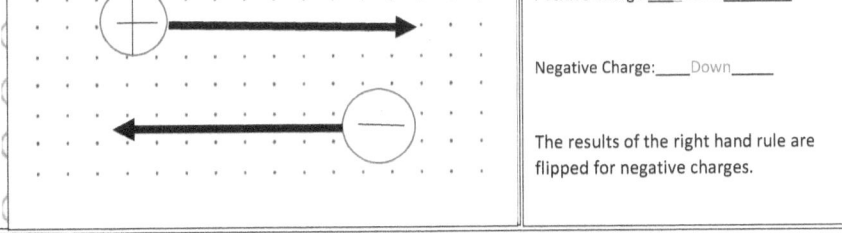

Positive Charge:___Down_____

Negative Charge:___Down____

The results of the right hand rule are flipped for negative charges.

Next Vloudel showed Zeno some pages on optics:

Convex and Concave Mirrors and Lenses

Optics

Light is an electromagnetic wave. It is composed of an electric and a magnetic field that oscillates at 90 degrees from each other.

The electric field = $\frac{Magnetic\ field}{\sqrt{\mu_0 e_0}}$

Velocity of light = $\frac{1}{\sqrt{\mu_0 e_0}}$

Mirrors

Concave mirrors:

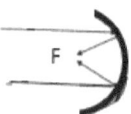

Converges light in front of mirror. Focal length is positive and real. The image can be positive real or negative virtual.

Convex mirrors:

Converges light behind mirror. Focal length is negative virtual. Image is always negative virtual.

Reflection of Light:

Light reflects off surfaces at the same angle as the incident angle. The angle is measured from the normal.

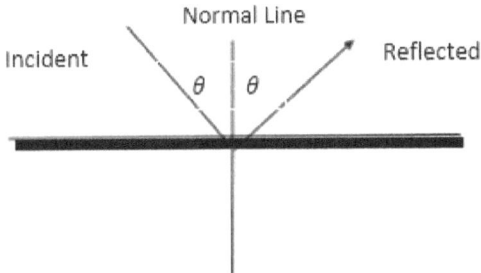

The equation for mirrors and lenses is:

$$\frac{1}{object's\ distance} + \frac{1}{image's\ distance} = \frac{1}{focal\ length}$$

The image and focal length can be negative or positive depending if they are virtual or real.

The focal length is where light from infinity converges.

Concave mirrors:

C is the center of curvature 2x the focal length

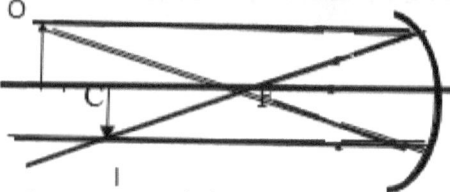

Image is real, inverted, and smaller.

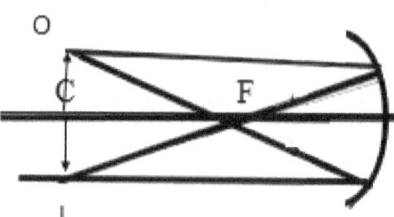

Image is inverted, real, and the same size.

The further away the image is from the mirror the bigger it is. In this case the image is just as far away as the object, so their sizes are the same.

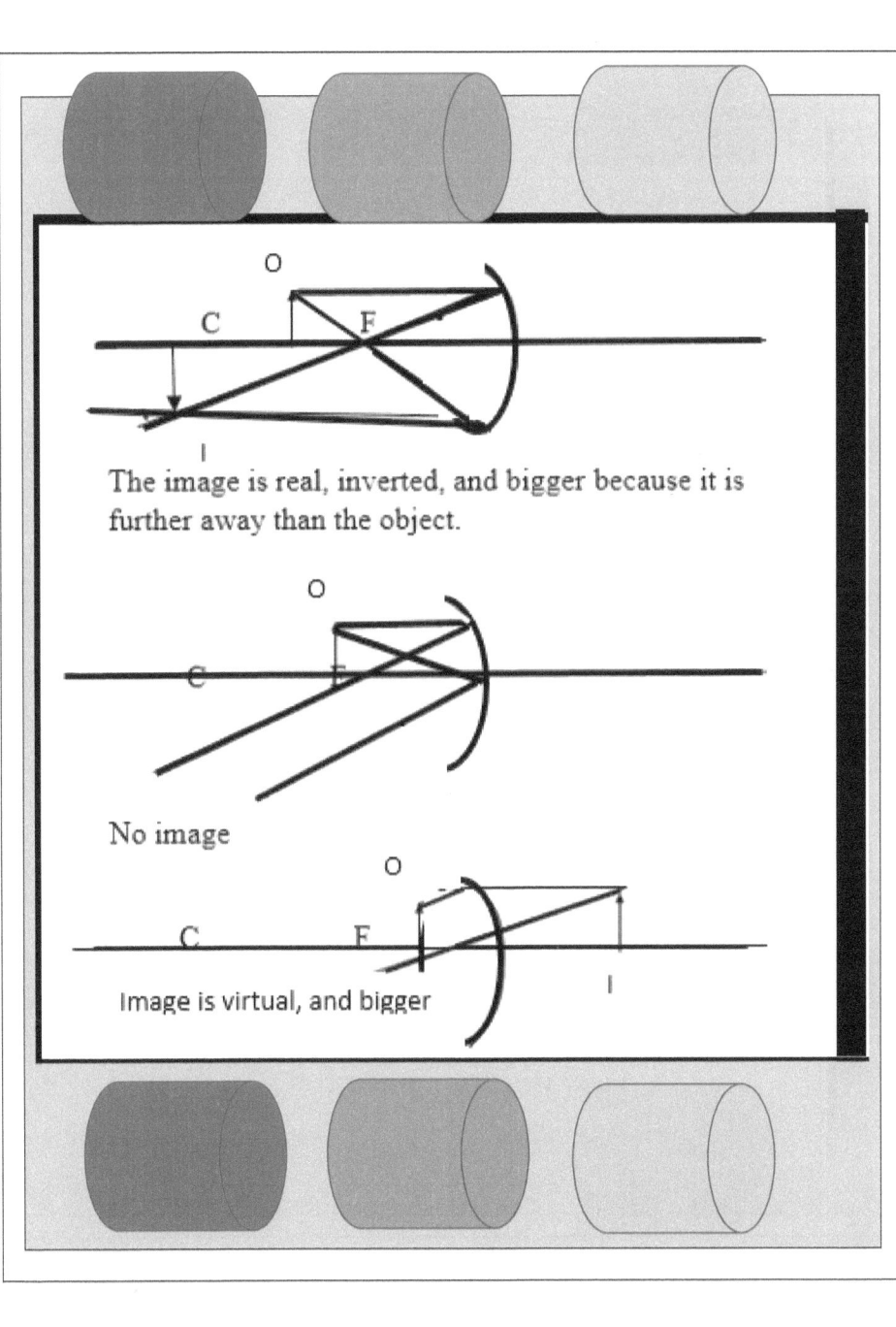

Convex mirrors:

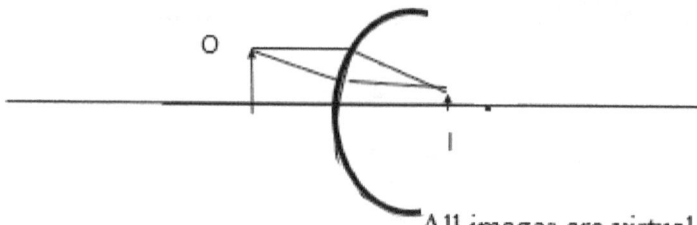

smaller
All images are virtual and

Lenses

Concave lenses:

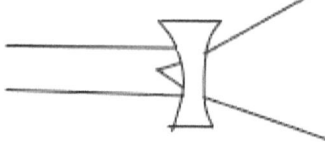

Focal length is negative virtual

Convex lenses:

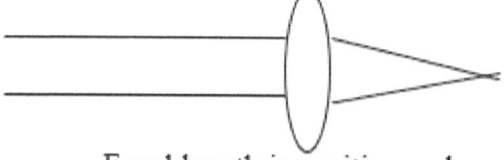

Focal length is positive real

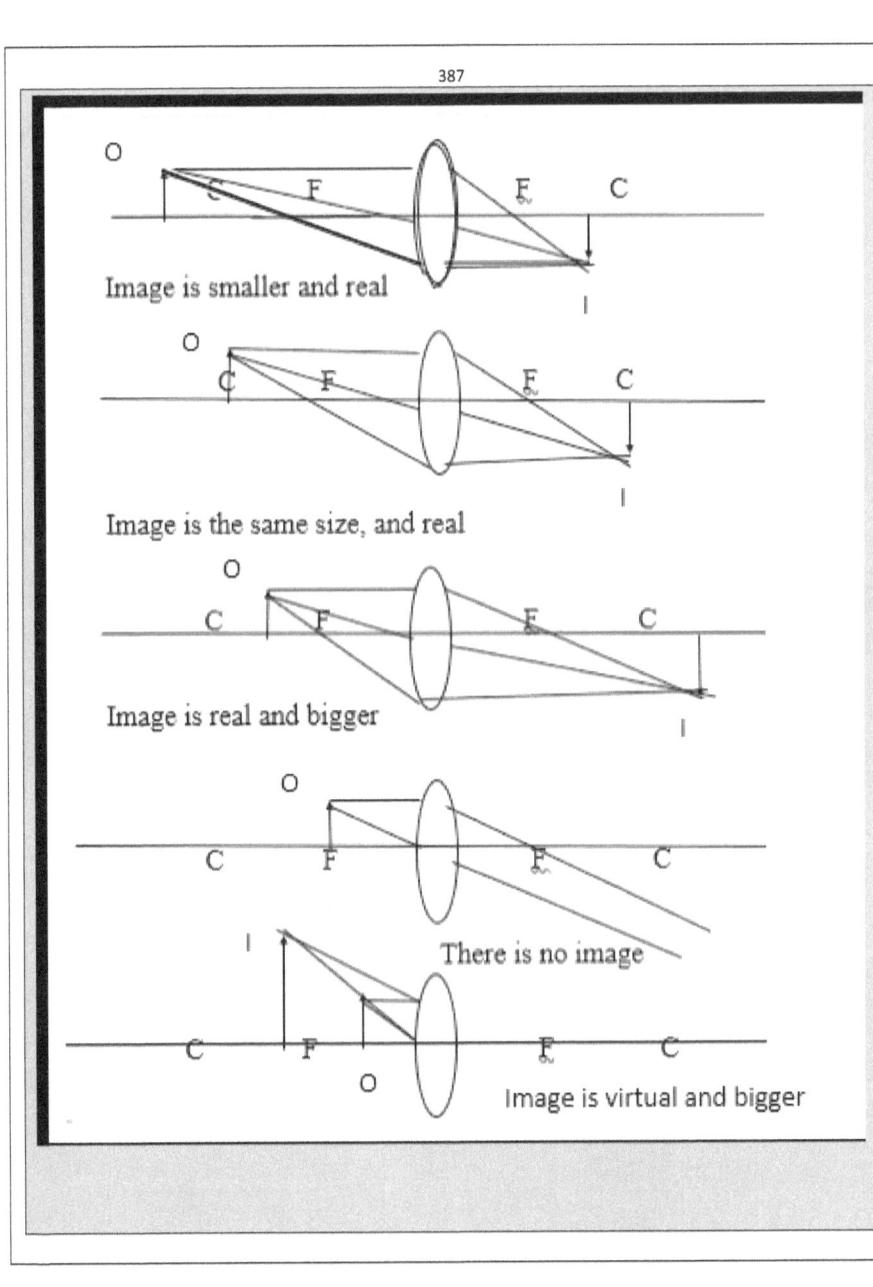

Image is virtual and bigger

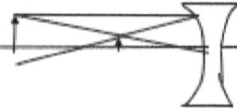

For concave lenses the image is always virtual and smaller.

Magnification of images = $\dfrac{-image\ location}{object\ location} = \dfrac{-Height\ of\ image}{Height\ of\ object}$

Magnification is positive if the image is virtual.

If the image is virtual it is right side up.

If the image is real, it is upside down, and its magnification is negative.

When light goes through two slits it causes an interference on a screen after the two slits.

Constructive interference causes the visible light patches that are followed by destructive interference that are the dark patches.

Constructive interference occurs at the following angles:

$Sin\theta = \frac{m\lambda}{D}$ where m = 0,1,2,3.....

Where D is the distance between the two slits and λ is the wavelength of the light.

Destructive interference occurs at the following angles:

$Sin\theta = (m + 0.5)\frac{\lambda}{D}$, where m = 0,1,2,3....

When light goes through a single slit, it also produces an interference. The location of its dark bars is at the following angles:

$Sin\theta = \frac{m\lambda}{w}$ where m=1,2,3...and w is the width of the slit.

When using a telescope, the minimum angle where it is possible to distinguish 2 points separately is:

$Sin\theta = \frac{1.22\lambda}{D}$,

Where D is the diameter of the telescope's aperture, and λ is the wavelength of the incoming light.

Waves are Everywhere

Explanation for Double Slit Experiment

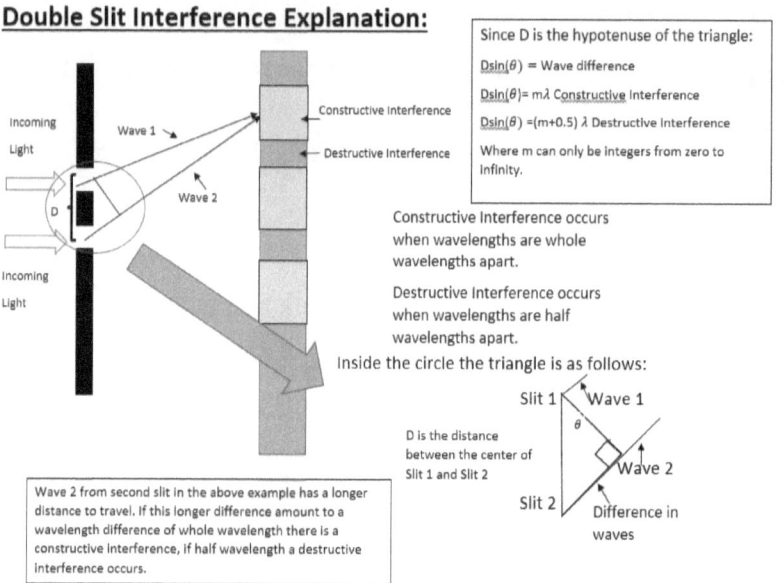

When the difference in wavelength is a whole number of wavelengths there is a constructive interference since the waves add up. When the difference in wavelength is a half wavelength the waves cancel and there is a destructive interference.

In this discussion you will answer the following questions with 5 sentences each, explaining your perspective on how the future of science will lead human civilization towards another revolution:

1) Explain how advances in science such as Biology could permit humans to live longer than 100 years, and how a biological process of hormones and chemicals is possibly preferred instead of electronic chip implants. Describe how all diseases may get a cure, and what would be the effect of longevity in human evolution. (At least 5 sentences)

2) Explain how a Biological Revolution could lead to a generation of humans that are smarter through a manipulation of the neurons, and chemicals that favor critical thinking in the brain. How would humans be different if the chemicals for intelligence were discovered and every human became

smarter than the smarted person on earth today. (At least 5 sentences)

3) Briefly summarize the effects of longevity and increase in intelligence that could derive from a Biological Revolution in the future. (At least 5 sentences)

Discussion 2:

1)

The universe contains 2 trillion galaxies, and each galaxy has an average of 100 to 200 billion stars, and each star is a sun. These are very big numbers which may lead one to question: "Does it even make sense based on pure reason to say that the earth is the only planet with life? If life was born by pure chance on earth, then why could it not be formed likewise elsewhere in the cosmos? The sun was formed in a cluster of stars, and over time this cluster expanded with stars moving away from each other. Today the sun is not part of a cluster and its sister stars are unknown. If the chemicals necessary for life on earth was present in our solar system, then why would it not be present in the sun's sister stars since they are formed together in a single cluster?

Did life began on earth or was brought to earth from elsewhere arriving on earth from the collision of a comet, or asteroid?

If there is life in other planets, and if this life is intelligent then why have these Aliens never contacted us? The earth exists from 4 billion years, and life began on earth 3.5 billion years ago in the form of bacteria.

Life then evolved into beings that we can see with our own eyes.

Life then moved from water to land.

Later came the Dinosaurs who lived on earth between 165 to 177 million of years ago.

Following the track of evolution the first human was born 300,000 years ago, and the first human civilization was formed 6,000 years ago.

Radio technology then only became available to us 100 years ago.

This shows that the percentage of the time in which earth's intelligent beings which are us, is capable of extraterrestrial contact with respect to the total geological time of earth is: $((100 \text{ years})/(4 \text{ billion years}))100 = 2.5 \times 10^{-6}$ %. This shows that the likelihood that an extraterrestrial civilization could enter in contact with us is extremely small. What if these extraterrestrials were alive when the earth only had dinosaurs, or bacteria, or other times earth's history, and now that we have the technology these extraterrestrials are nowhere to be found? This shows that as we humans try contact with them, these beings of other planets could also be like dinosaurs, or bacteria, or even intelligent but without radio technology. In other words, to find life in other planets is very hard. These beings could even be very developed enough to hide from us, to prevent us from ever finding them. If there are only 9 planets with life in our galaxy, and knowing that our galaxy has 200 billion stars, finding these 9 planets among hundreds of billion others is a colossal work that will take an extremely long time if possible. In other words, most intelligent beings in the universe, if they truly exist outside the earth, all have a feeling of loneliness in this silent cosmos.

In this discussion you will answer all the questions with a reply of 500 words in this text and add a conclusion which should be at least 7 sentences long:

Discussion 3:

Here you will answer the following questions about wave, energy, sound, and light. You will be graded not on how accurate your answers are but on how well you explain your ideas.

1) If something is said to be small and the other to be big, it is important to have a reference. The Moon is small relative to the Earth but big relative to space shuttle. The Sun is big relative to Earth but small relative to an entire Galaxy. An atom is big relative to an Electron but small relative to a chair. Explain why is it important to have a point of reference when stating how big something is. (At least 5 sentences)

2) All matter are made of waves and all waves are formed from a combination of Sine and Cosine. Explain why is it important to have a good understanding of Trigonometry in trying to understand the universe. The entire universe is a collection of wave functions and that gives structure to reality. (At least 5 sentences)

3) If Energy is also a wave, and Energy is conserved, explain how the universe has a limited amount of Energy since the beginning of the universe. Ever since the Big Bang the amount of Energy and matter is the same, and the only difference is that the universe expanded and this Energy and matter is more spread. Explain why Energy has to be conserved. (At least 5 sentences)

Discussion 4:

Brain waves can now be transferred into electrical signals and stored in a computer that is able to match the waves to feelings in the form of images and even sound. That means that now that Artificial Intelligence is able to read people's mind by matching brain waves with thoughts, telepathy can be possible as long as people have their minds connected to each other through the Internet. The computer was taught to relate thoughts in the form of waves and to generate images and sounds that are a perfect match to what is really going on in a person's mind. Does that mean that the sense of privacy is gone? Would you connect your mind to the Internet or download your thoughts in a Pen Drive? Would you download your entire mind into Google Drive? Would you access the Internet with just a mere thought without the need to type on Google Search or speak anything, but just with a thought?

Thoughts downloaded into a PenDrive

1) What is your opinion on having humans connect their mind to computers?

2) How could such scientific revolution be beneficial and harmful to people? How different would human civilization, sense of self, privacy, and overall knowledge be affected by having everyone be connected to computers and having their thoughts accessed through the internet? How would life be different when everyone is connected to each other and are able to transfer information and thoughts by using the internet with only their thoughts? (At least 200 words)

While using computers to understand the human brain:

The question is: Can you live without your phone or the Internet? The future is filled with possibilities and facts.

If the answer is no to the above question then that means that you are already a step ahead into a complete mind to Internet
Will the brain and the Internet become one?

Are we destined to become one single entity? anymore. Is it likely that this brain to machine fusion will happen, or will it be ethically wrong and never accepted in the society? Write down your thoughts on thi

Name_____Period_____

Circuits Review

Table of Resistivities:

1...Answer the following questions using the table on the right:

Silver: $1.59 \times 10^{-8} \, \omega m$

Copper: $1.68 \times 10^{-8} \, \omega m$

Aluminum: $2.65 \times 10^{-8} \, \omega m$

Wow! Let us venture!

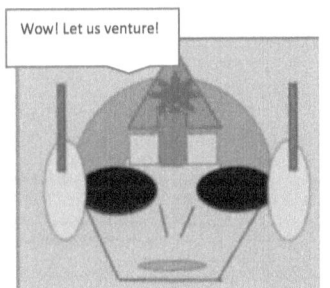

First you must find the Resistance of each resistor.

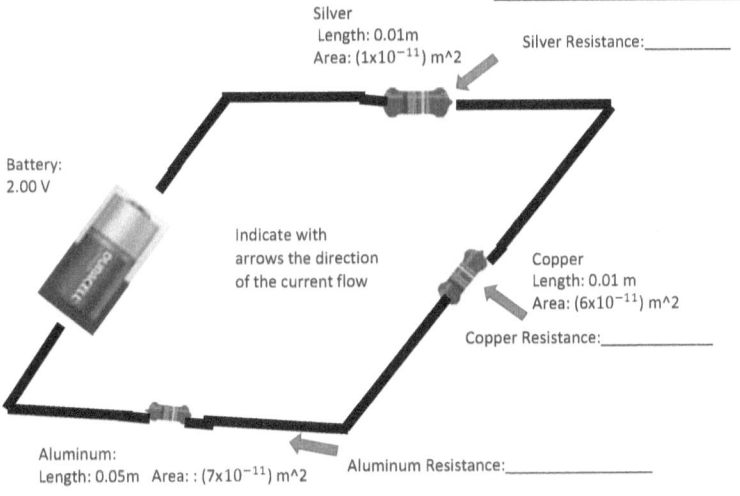

Silver
Length: 0.01m
Area: (1×10^{-11}) m^2

Silver Resistance:_____

Battery:
2.00 V

Indicate with arrows the direction of the current flow

Copper
Length: 0.01 m
Area: (6×10^{-11}) m^2

Copper Resistance:_____

Aluminum:
Length: 0.05m   Area: : (7×10^{-11}) m^2

Aluminum Resistance:_____

Fill the Table below for the circuit in Page 1:

Resistor	Voltage	Current	Resistor	Power
Total				

If the Resistors were Bulbs which one will shine brightest?

_____

If the Resistors were Bulbs which one will be the dimmest?

_____

State what happens to resistance when the following changes are made to the resistor:

A....Length doubled:_____
B....Length halved:_____
C....Area doubled:_____
D...Area cut to a third:_____

What is the Total Resistance of a Parallel Circuit with the following Resistors:

12 ohms, 13 ohms, 22 ohms, 23 ohms, 56 ohms, 57 ohms, and 89 ohms:

_____ Ohms

Using the following Circuit fill the table:

Parallel Circuit

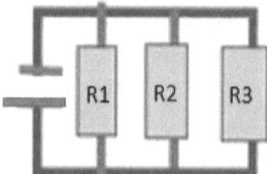

Resistor	Voltage	Current	Resistance	Power
			10.0ω	
			20.0ω	
			14.0 ω	
Total	9.00 V			

If the Resistors were Bulbs which one will shine brightest?

_____

If the Resistors were Bulbs which one will be the dimmest?

_____

Using the following Circuit fill the table:

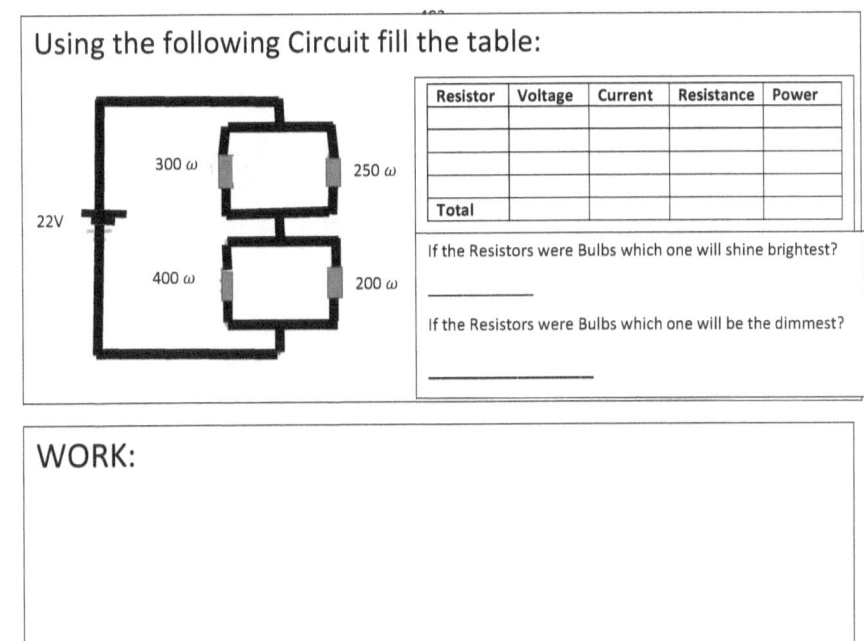

Resistor	Voltage	Current	Resistance	Power
Total				

If the Resistors were Bulbs which one will shine brightest?

_____

If the Resistors were Bulbs which one will be the dimmest?

_____

WORK:

Using the following Circuit fill the table:

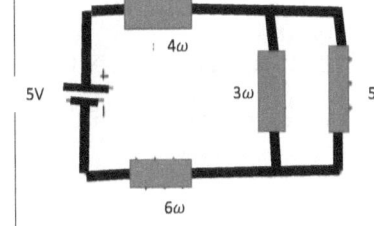

Resistor	Voltage	Current	Resistance	Power
Total				

If the Resistors were Bulbs which one will shine brightest?

_____

If the Resistors were Bulbs which one will be the dimmest?

_____

Name_____ Period_____

Right Hand Rule Review 2

1…What is the Electric Field at a Distance of 8.9mm from a Particle with charge -7.8cC?

2…Does the Field in Problem 1 point away or towards the Particle?_____

3…If now a Particle with 6.6cC of charge is placed at that same location from the -7.8 cC Particle, what is the Magnitude of the Force between them?

4…Is the Force Attractive of Repulsive?

_____

5…A loop is placed in a Magnetic Field with current in the direction shown with the arrows.

The Physics of Electric Motors

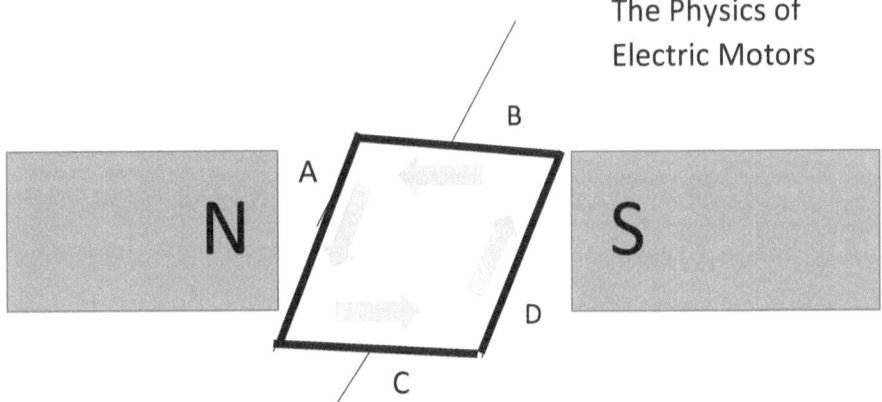

This set up will cause the loop to spin. State the direction of the Force on the side A and Side D of the Loop:

Side A_____

Side D _____

To keep the loop rotating with full revolutions, the directions of the Currents are inverted many times.

6…You now insert a magnet with its North side into a loop of wire.

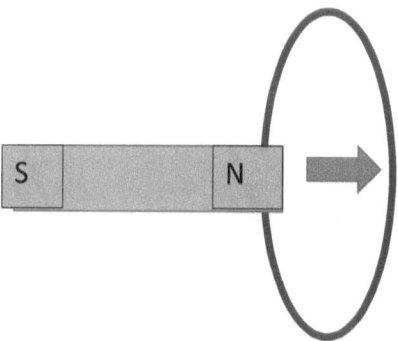

Draw arrows around the loop indicating which way will the Current flow in the wire?

The universe is composed of many loops of Electric and Magnetic Fields.

Draw arrows around the loop indicating which way will the current flow in the wire if now you are pulling the magnet out of the loop with its N side facing the loop:

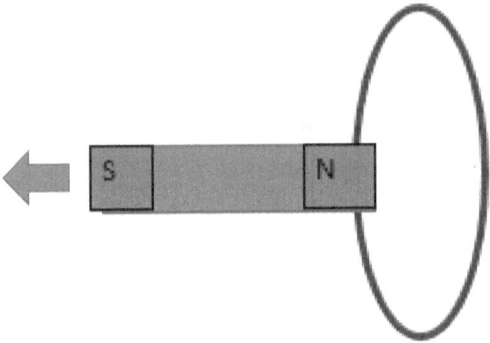

Write which of the Maxwell's Equations describes this phenomenon:

_____

 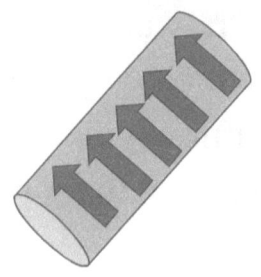

 The Solenoid on the right has current flowing in the direction shown. Indicate with an arrow the direction of the Magnetic Field it generates.

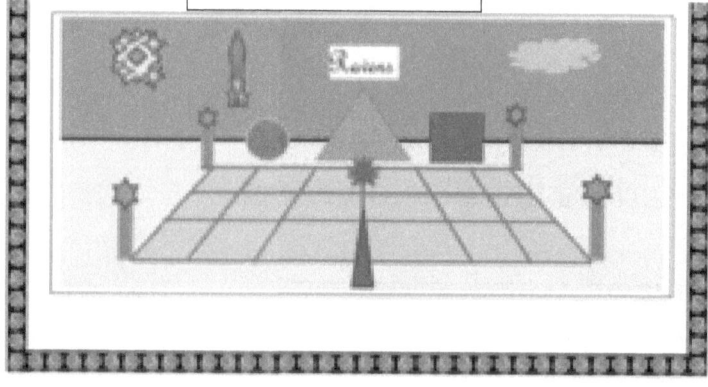

Everything is well measured and calculated

The wire below has current in the direction shown inside a region with a Magnetic Field as pictured. Draw an arrow for the Force that this wire will experience.

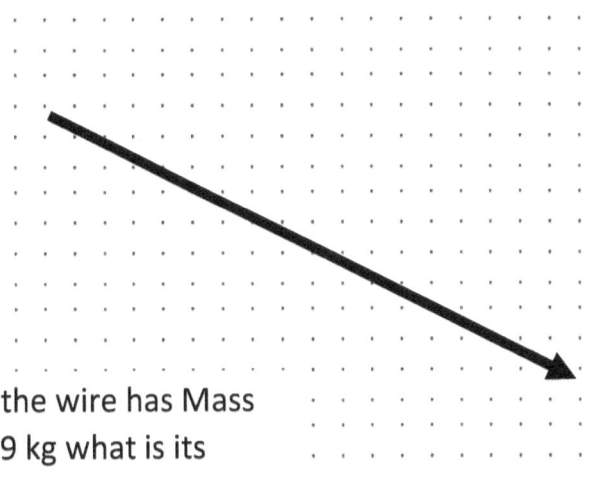

For this wire:

I = 2.0 A

L = 3.0 m

B = 4.0 T

What is the Force?
_____

If the wire has Mass 8.9 kg what is its initial acceleration?
_____

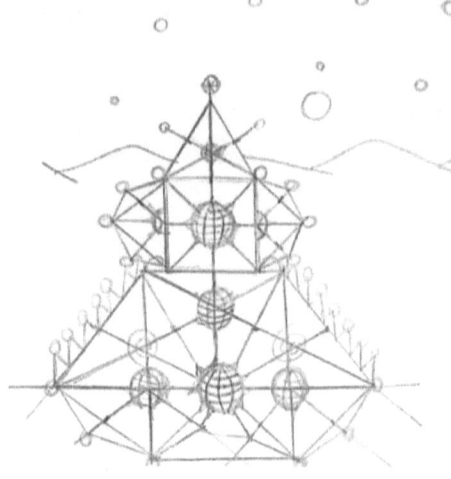

Mathematics and Geometry are in the Foundation of all the Laws and Constants found throughout the universe.

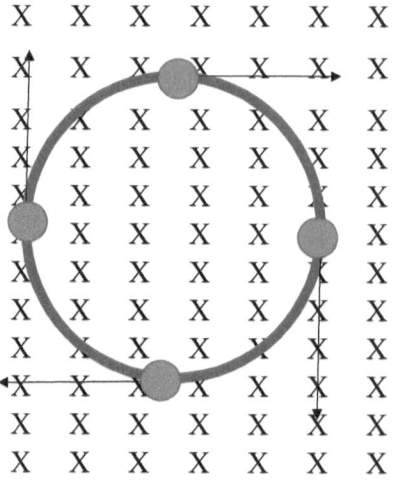

7...A Particle moves in a circle shown due to a Magnetic Field as demonstrated. Using the right-hand rule and taking note that the Force is towards the center of the circle, is the Particle positive of negative?

_____

8...Using the Equation:

Force = QvB

Where Q is the Charge, v the speed, and B the Magnetic Field, find the Force on the Particle if the Radius of the circle is 0.8mm and its mass is 7.8×10^{-7}kg, charge 6.7mC, and Magnetic Field 8.9T?

Force:_____

What is the Speed in which the Particle moves around the circle?

Speed:_____

> It may help finding the speed first and then the Force. Set QVB equal to Centripetal Force.

9...Explain which pair of wires will there be a force of attraction between the wires?

10...Use dots and Xs to describe the Magnetic Field around the following wires:

Please write which of Maxwell's Equations demonstrates Divergence:

_____

Please write which of Maxwell's Equations demonstrates that magnetism can only curl and never diverge:

_____

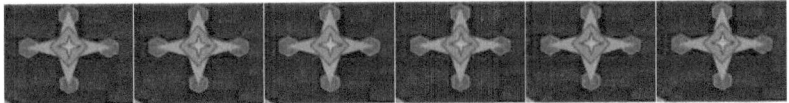

11....Find the Magnetic Field at the following Distances from a wire carrying 46mA of current:

Distance	Magnetic Field
5.6mm	
5.6cm	
5.6km	
5.6μm	
5.6m	

12....Find the Magnetic Field at the following Distances from a wire carrying 5.6cA of current:

Distance	Magnetic Field
5.6mm	
5.6cm	
5.6km	
5.6μm	
5.6m	

Name_____ Period___

Right Hand Rule Review

1...Draw the Electric Field Lines between the Positive and Negative Charges:

 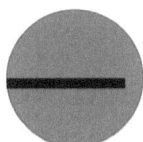

2...Draw the Electric Field Lines around a Positive Charge:

3…Draw Electric Field Lines around a Negative Charge:

4…Draw Electric Field Lines between the two Negative Charges:

5...Draw Electric Field Lines between two Positive Charges:

6...If the Particle on problem 5 on the left has a charge of $+6.7\mu C$ and the Particle on the right $+7.8\mu C$, while their Distance is 3.4mm, what is the Force between them?

Don't forget to convert the units.

7… If the Particle on problem 1 on the left has a charge of $+5.7mC$ and the Particle on the right $-9.8mC$, while their Distance is 6.4cm, what is the Force between them?

Keep an eye on these units.

8…State whether the Forces are Attractive of Repulsive:

Problem 6:_____

Problem 7:_____

Be aware of the signs of the Charges.

9…If there is a Particle with charge 8.0C, what is the Electric Field 5.6 m away from it?

 5.6 m

10…Will the Electric Field on problem 9 be towards the Particle or away?_____.

There are particles everywhere in the cosmos causing it to move and become the reality that we perceive.

420

11…Draw a Force Vector for the Wires carrying Current in the following directions

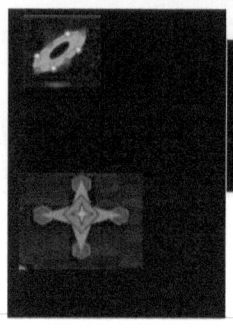

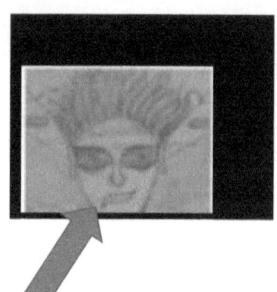

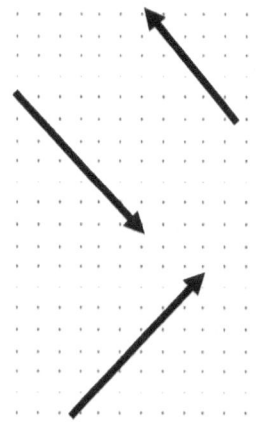

Space is filled with Electric and Magnetic Fields:

x x x x x x x   x x x x x x x   x x x x x x x
x x x x x x x   x x x x x x x   x x x x x x x
x x x x x x x   x x x x x x x   x x x x x x x
x x x x x x x   x x x x x x x   x x x x x x x
x x x x x x x   x x x x x x x   x x x x x x x
x x x x x x x   x x x x x x x   x x x x x x x
x x x x x x x   x x x x x x x   x x x x x x x
x x x x x x x   x x x x x x x   x x x x x x x
x x x x x x x   x x x x x x x   x x x x x x x
x x x x x x x   x x x x x x x   x x x x x x x
x x x x x x x   x x x x x x x   x x x x x x x

12…Label the Spectrum of Light:

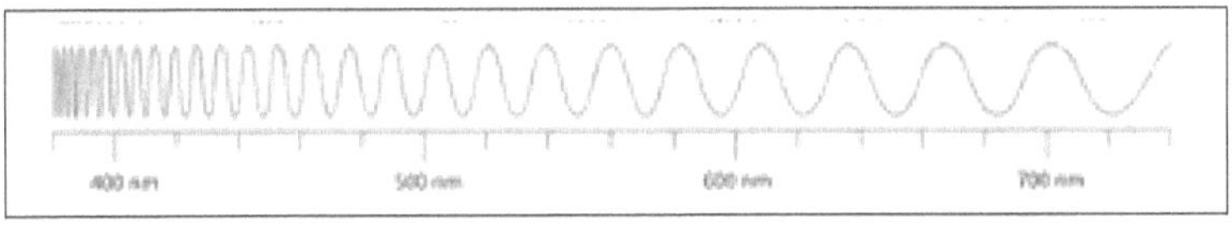

____ ____ ____ ____ ____ ____ ____ ____

400 nm   500 nm   600 nm   700 nm

13…Label the Colors of Light from Most Energetic to Least Energetic:

Most Energetic:_____

_____

_____

_____

_____

_____

Least Energetic:_____

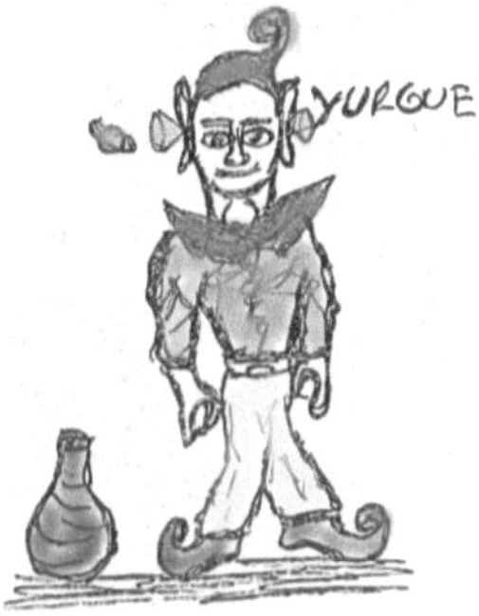

14...Draw an arrow for the direction of the Magnetic Field produced in these Solenoids (Coils) with current in the direction shown:

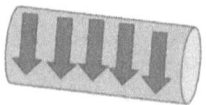

15... What is the Magnetic Field from a wire carrying 7.8mA of Current at a distance 5.6 cm away?

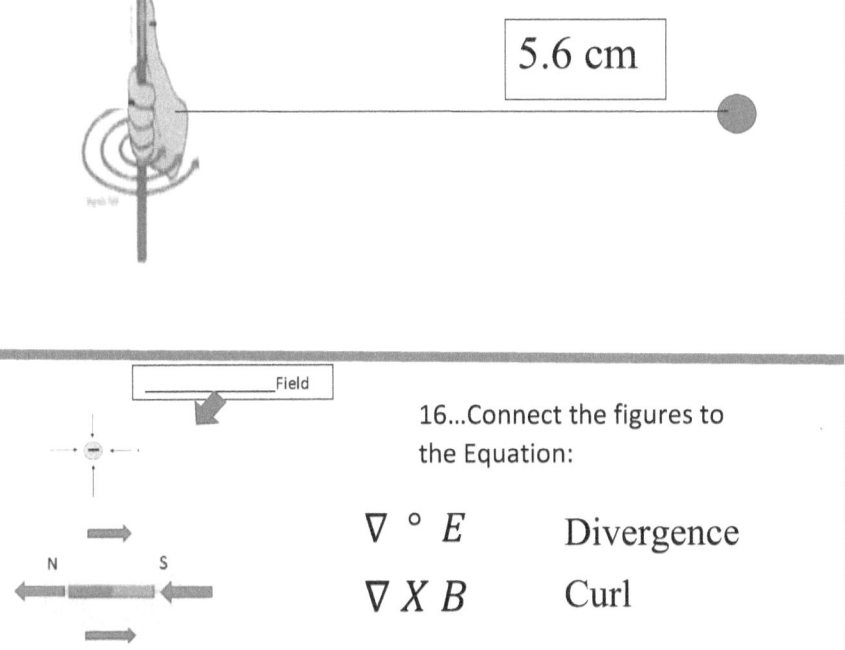

5.6 cm

_____ Field

16... Connect the figures to the Equation:

$\nabla \circ E$ Divergence

$\nabla \times B$ Curl

Also state what kind of Field are the figures representing.

_____ Field

17...Explain what each equation represents:

$\nabla \cdot \mathbf{E} = \dfrac{\rho}{\varepsilon_0}$

$\nabla \cdot \mathbf{B} = 0$

$\nabla \times \mathbf{E} = -\dfrac{\partial \mathbf{B}}{\partial t}$

$\nabla \times \mathbf{B} = \mu_0 \mathbf{j} + \dfrac{1}{c^2}\dfrac{\partial \mathbf{E}}{\partial t}$

Light is an Electromagnetic Wave that spins and twists as it travels through space. Many things spin in space. Stars, Planets, Particles, and Galaxies. Spinning is the way the cosmos dance in this Universal Flow.

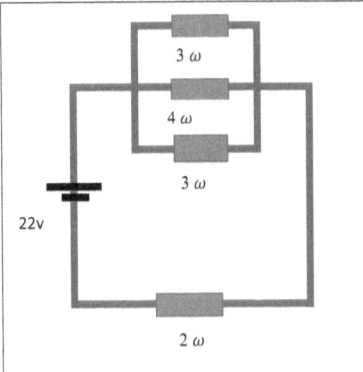

Resistor	Voltage	Current	Resistance	Power
Total				

If the Resistors were Bulbs which one will shine brightest?

_____

If the Resistors were Bulbs which one will be the dimmest?

_____

WORK:

Using the following Circuit fill the table:

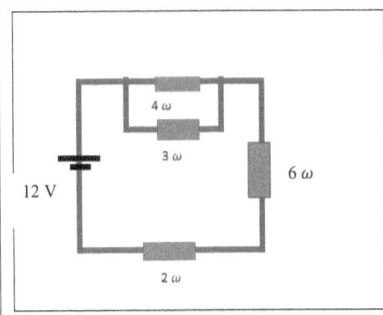

Resistor	Voltage	Current	Resistance	Power
Total				

If the Resistors were Bulbs which one will shine brightest?

_____

If the Resistors were Bulbs which one will be the dimmest?

Name_____Period_____

Circuits Test

Table of Resistivities:

1...Answer the following questions using the table on the right:

VVVVVVVVVVVVVVVVVVVVVVVVV

Silver: $1.59 \times 10^{-8}\ \omega m$
Copper: $1.68 \times 10^{-8}\ \omega m$
Aluminum: $2.65 \times 10^{-8}\ \omega m$

Wow! Let us venture!

First you must find the Resistance of each resistor.

Silver
Length: 0.02m
Area: (2×10^{-11}) m^2

Silver Resistance:_____

Battery:
2.00 V

Indicate with arrows the direction of the current flow

Copper
Length: 0.03 m
Area: (3×10^{-11}) m^2

Copper Resistance:_____

Aluminum:
Length: 0.02m   Area: : (2×10^{-11}) m^2

Aluminum Resistance:_____

Fill the Table below for the circuit in Page 1:

Resistor	Voltage	Current	Resistor	Power
Total				

If the Resistors were Bulbs which one will shine brightest?

_____

If the Resistors were Bulbs which one will be the dimmest?

_____

State what happens to resistance when the following changes are made to the resistor:

A….Length triples:_____
B….Length halved:_____
C….Area tripled:_____
D…Area cut to a fourth:_____

What is the Total Resistance of a Parallel Circuit with the following Resistors:

10 ohms, 10 ohms, 20 ohms, 23 ohms, 50 ohms, 57 ohms, and 80 ohms:

_____ Ohms

Using the following Circuit fill the table:

Parallel Circuit

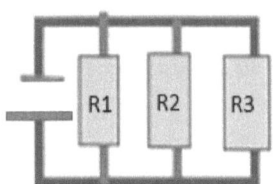

Resistor	Voltage	Current	Resistance	Power
			1.0ω	
			2.0ω	
			4.0 ω	
Total		9.00 V		

If the Resistors were Bulbs which one will shine brightest?

_____

If the Resistors were Bulbs which one will be the dimmest?

_____

Using the following Circuit fill the table:

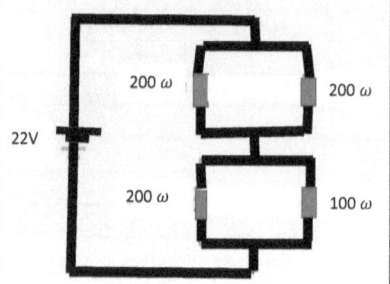

Resistor	Voltage	Current	Resistance	Power
Total				

If the Resistors were Bulbs which one will shine brightest?

_____

If the Resistors were Bulbs which one will be the dimmest?

_____

WORK:

Using the following Circuit fill the table:

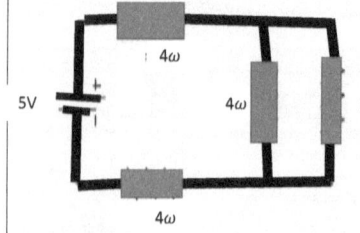

Resistor	Voltage	Current	Resistance	Power
Total				

If the Resistors were Bulbs which one will shine brightest?

_____

If the Resistors were Bulbs which one will be the dimmest?

_____

Name _____ Period _____

Right Hand Rule Review 1

1…What is the Electric Field at a Distance of 9.9mm from a Particle with charge -3.8C?

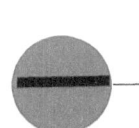 ———————— 9.9 mm ————

2…Does the Field in Problem 1 point away or towards the Particle? _____

3…If now a Particle with 6.6cC of charge is placed at that same location from the -3.8C Particle, what is the Magnitude of the Force between them?

4...Is the Force Attractive of Repulsive?

_____

5...A loop is placed in a Magnetic Field with current in the direction shown with the arrows.

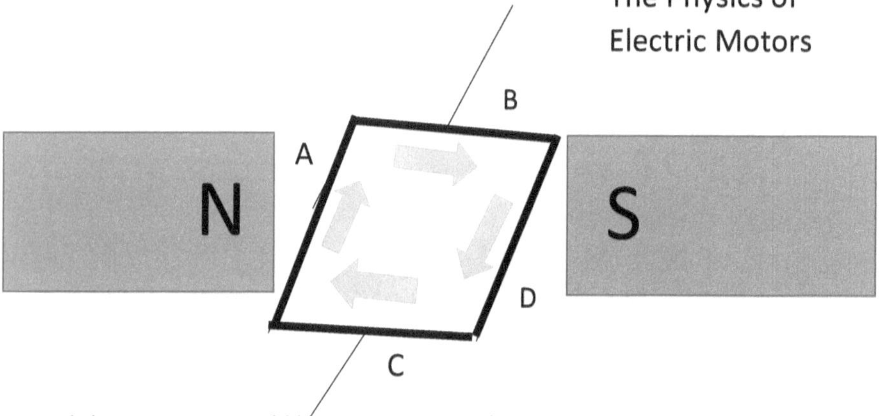

The Physics of Electric Motors

This set up will cause the loop to spin. State the direction of the Force on the side A and Side D of the Loop:

Side A_____

Side D _____

To keep the loop rotating with full revolutions, the directions of the Currents are inverted many times.

6…You now insert a magnet with its South side into a loop of wire.

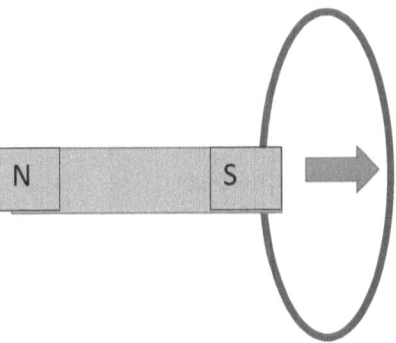

Draw arrows around the loop indicating which way will the Current flow in the wire?

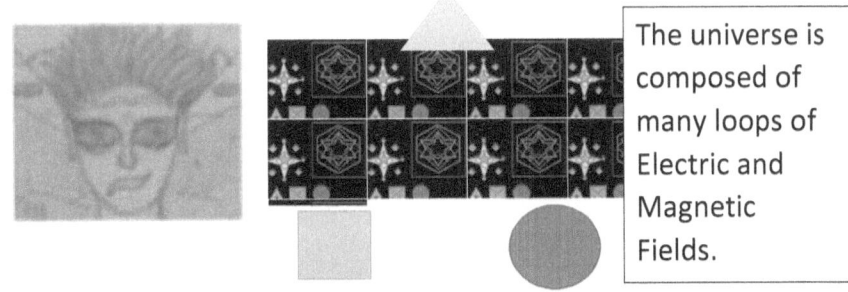

The universe is composed of many loops of Electric and Magnetic Fields.

7...Draw arrows around the loop indicating which way will the current flow in the wire if now you are pulling the magnet out of the loop with its N side facing the loop:

Write which of the Maxwell's Equations describes this phenomenon:

_____

8…Explain which pair of wires will there be a force of attraction between the wires?

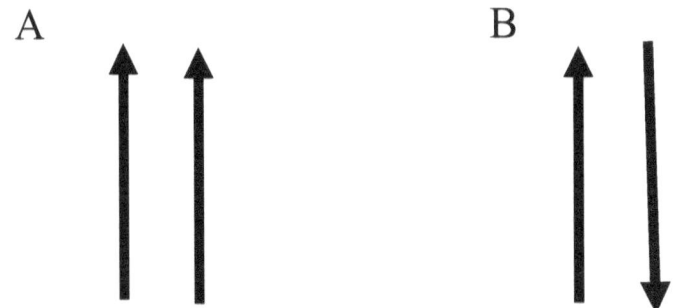

9…Use dots and Xs to describe the Magnetic Field around the following wires:

10…Draw an arrow for the direction of the Magnetic Field produced in these Solenoids (Coils) with current in the direction shown:

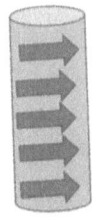

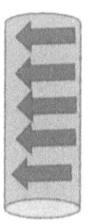

11... What is the Magnetic Field from a wire carrying 2.8A of Current at a distance 2.6 m away?

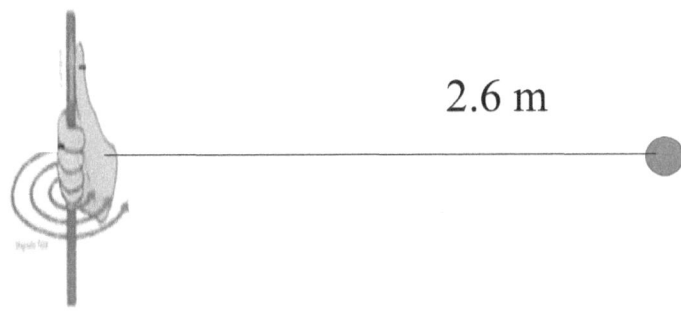

2.6 m

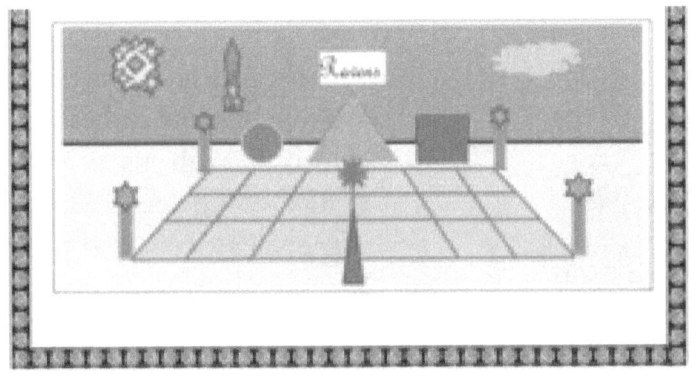

12...Explain what each equation represents:

$$\nabla \cdot \mathbf{E} = \frac{\rho}{\varepsilon_0}$$

$$\nabla \cdot \mathbf{B} = 0$$

$$\nabla \times \mathbf{E} = -\frac{\partial \mathbf{B}}{\partial t}$$

$$\nabla \times \mathbf{B} = \mu_0 \mathbf{j} + \frac{1}{c^2}\frac{\partial \mathbf{E}}{\partial t}$$

Light is an Electromagnetic Wave that spins and twists as it travels through space. Many things spin in space. Stars, Planets, Particles, and Galaxies. Spinning is the way the cosmos dance in this Universal Flow.

13...If there is a Particle with charge 1.0C, what is the Electric Field 2.6 m away from it?

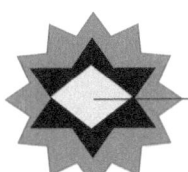   2.6 m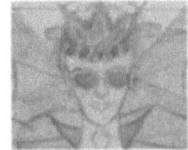

14...Will the Electric Field on problem 13 be towards the Particle or away?_____.

There are particles everywhere in the cosmos causing it to move and become the reality that we perceive.

15...Draw a Force Vector for the Wires carrying Current in the following directions

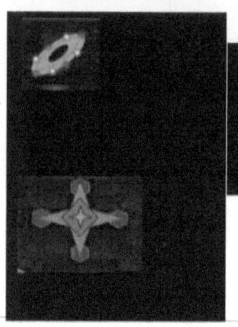

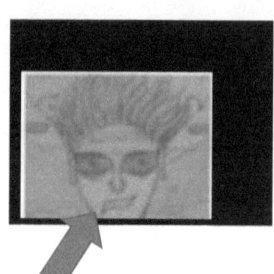

Space is filled with Electric and Magnetic Fields:

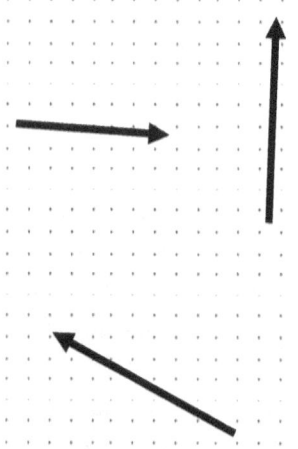

The wire below has current in the direction shown inside a region with a Magnetic Field as pictured. Draw an arrow for the Force that this wire will experience.

For this wire:

I = 3.0 A

L = 6.0 m

B = 7.0 T

What is the Force?
_____

If the wire has Mass 1.9 kg what is its initial acceleration?
_____

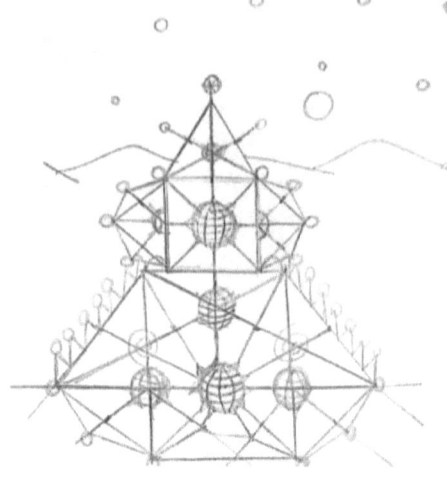

Mathematics and Geometry are in the Foundation of all the Laws and Constants found throughout the universe.

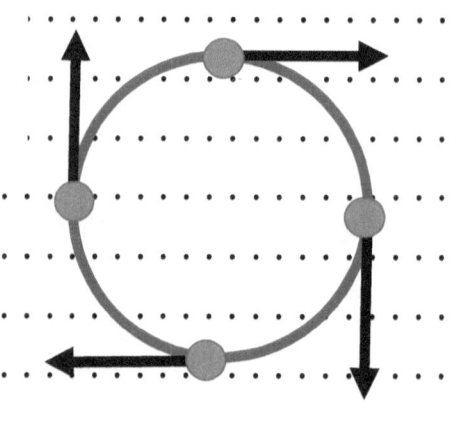

17...A Particle moves in a circle shown due to a Magnetic Field as demonstrated. Using the right-hand rule and taking note that the Force is towards the center of the circle, is the Particle positive of negative?

_____

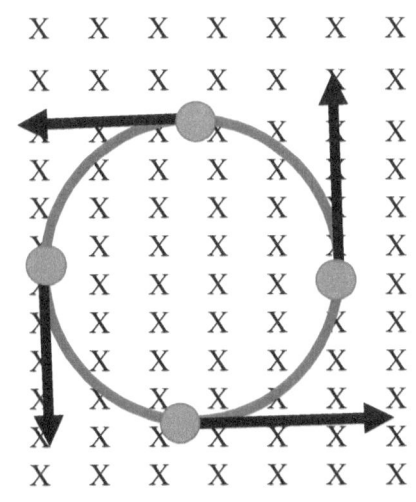

18...A Particle moves in a circle shown due to a Magnetic Field as demonstrated. Using the right-hand rule and taking note that the Force is towards the center of the circle, is the Particle positive of negative?

_____

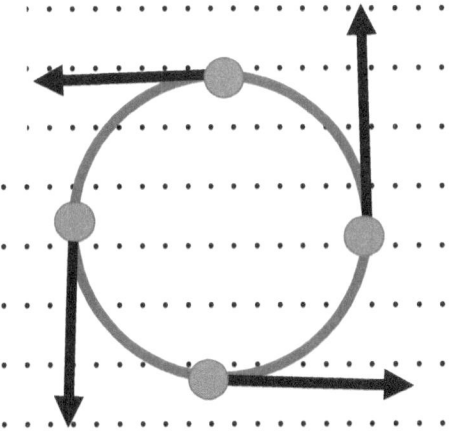

19...A Particle moves in a circle shown due to a Magnetic Field as demonstrated. Using the right-hand rule and taking note that the Force is towards the center of the circle, is the Particle positive of negative?

_____

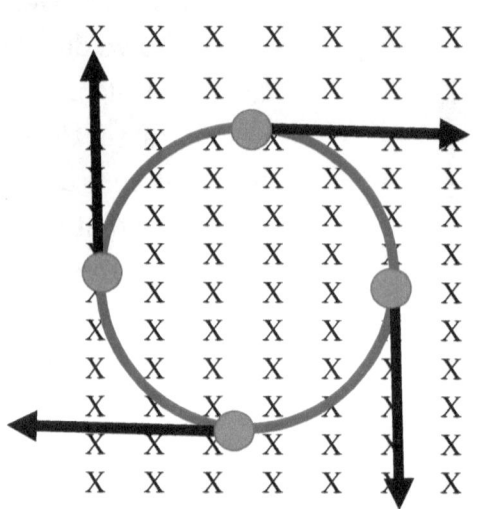

20…A Particle moves in a circle shown due to a Magnetic Field as demonstrated. Using the right-hand rule and taking note that the Force is towards the center of the circle, is the Particle positive of negative?

_____

21…Label the Colors of Light from Most Energetic to Least Energetic:

Most Energetic: _____

_____

_____

_____

_____

_____

Least Energetic: _____

Name_____Period_____

Circuits 2

Fill out the Table below:

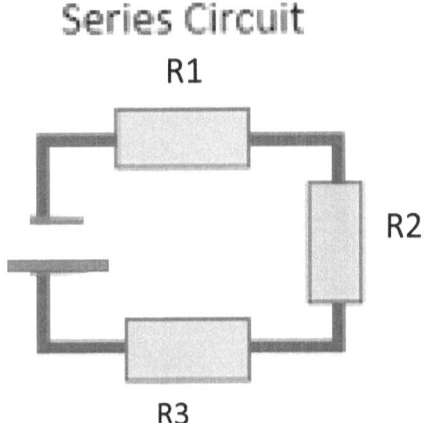

Resistor	Voltage	Current	Resistance	Power
1			19.0 ohms	
2			20.0 ohms	
3			15.0 ohms	
Total	80.0 V			

If Resistors were bulbs which one will shine brightest and which one dimmest?

Brightest_____Dimmest_____

Second Table

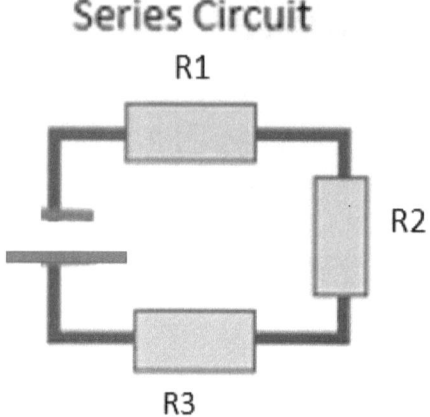

Series Circuit

Resistor	Voltage	Current	Resistance	Power
1		89.0A	1.0 ohms	
2			2.0 ohms	
3				
Total			5.0 ohms	

If Resistors were bulbs which one will shine brightest and which one dimmest?

Brightest_____ Dimmest_____

3-Third Table

Parallel Circuit

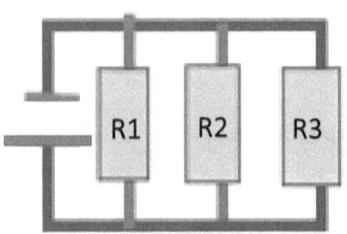

Resistor	Voltage	Current	Resistance	Power
1			7.0 ohms	
2			2.0 ohms	
3			10 ohms	
Total	3.0 V			

If Resistors were bulbs which one will shine brightest and which one dimmest?

Brightest_____Dimmest_____

4-Fourth Table

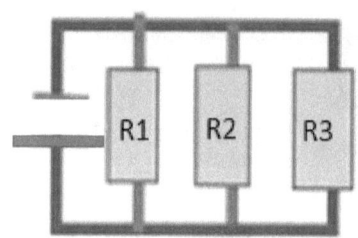

Resistor	Voltage	Current	Resistance	Power
1	50.0 V		10.0 ohms	
2			12.0 ohms	
3			11.5 ohms	
Total				

If Resistors were bulbs which one will shine brightest and which one dimmest?

Brightest_____Dimmest_____

Using the following Circuit fill the table:

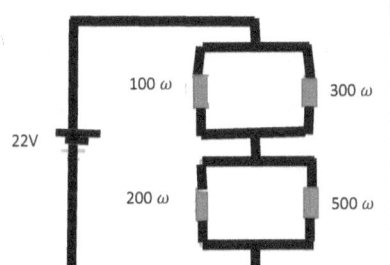

Resistor	Voltage	Current	Resistance	Power
Total				

If the Resistors were Bulbs which one will shine brightest?
_____

If the Resistors were Bulbs which one will be the dimmest?
_____

WORK:

Using the following Circuit fill the table:

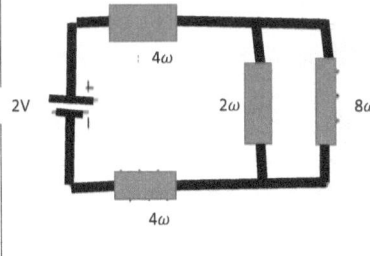

Resistor	Voltage	Current	Resistance	Power
Total				

If the Resistors were Bulbs which one will shine brightest?
_____

If the Resistors were Bulbs which one will be the dimmest?
_____

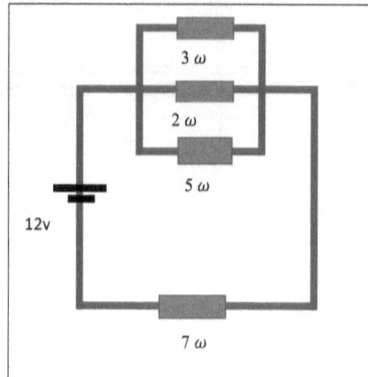

Resistor	Voltage	Current	Resistance	Power
Total				

If the Resistors were Bulbs which one will shine brightest?

_____

If the Resistors were Bulbs which one will be the dimmest?

_____

WORK:

Using the following Circuit fill the table:

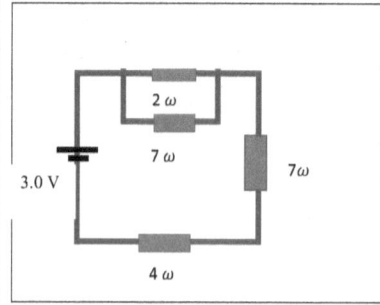

Resistor	Voltage	Current	Resistance	Power
Total				

If the Resistors were Bulbs which one will shine brightest?

_____

If the Resistors were Bulbs which one will be the dimmest?

Name_____Period_____

Circuits Review 3

Table of Resistivities:

1...Answer the following questions using the table on the right:

VVVVVVVVVVVVVVVVVVVVVVVVVV

Silver: $1.59 \times 10^{-8} \, \omega m$

Copper: $1.68 \times 10^{-8} \, \omega m$

Aluminum: $2.65 \times 10^{-8} \, \omega m$

Wow! Let us venture!

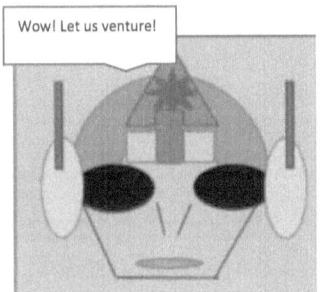

First you must find the Resistance of each resistor.

Silver
Length: 0.09m
Area: (1×10^{-11}) m^2

Silver Resistance:_____

Battery: 2.00 V

Indicate with arrows the direction of the current flow

Copper
Length: 0.01 m
Area: (1×10^{-11}) m^2

Copper Resistance:_____

Aluminum:
Length: 0.05m   Area: : (8×10^{-11}) m^2

Aluminum Resistance:_____

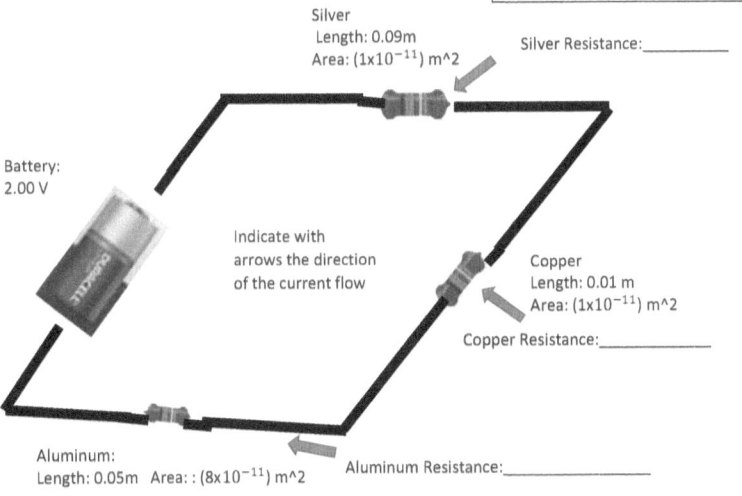

Fill the Table below for the circuit in Page 1:

Resistor	Voltage	Current	Resistor	Power
Total				

If the Resistors were Bulbs which one will shine brightest?

_____

If the Resistors were Bulbs which one will be the dimmest?

_____

State what happens to resistance when the following changes are made to the resistor:

A….Length quadruples:_____
B…Length cut to third:_____
C….Area doubled:_____
D…Area cut to a third:_____

What is the Total Resistance of a Parallel Circuit with the following Resistors:

11 ohms, 12 ohms, 23 ohms, 25 ohms, 59 ohms, 77 ohms, and 30 ohms:

_____ Ohms

Using the following Circuit fill the table:

Parallel Circuit

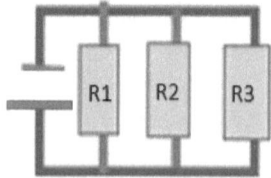

Resistor	Voltage	Current	Resistance	Power
			2.0ω	
			3.0ω	
			4.0 ω	
Total	3.00 V			

If the Resistors were Bulbs which one will shine brightest?

_____

If the Resistors were Bulbs which one will be the dimmest?

_____

Using the following Circuit fill the table:

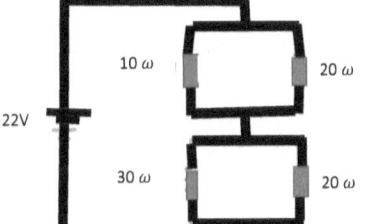

Resistor	Voltage	Current	Resistance	Power
1			10 ω	
2			20 ω	
3			30 ω	
4			20 ω	
Total				

If the Resistors were Bulbs which one will shine brightest?

_____

If the Resistors were Bulbs which one will be the dimmest?

_____

WORK:

Using the following Circuit fill the table:

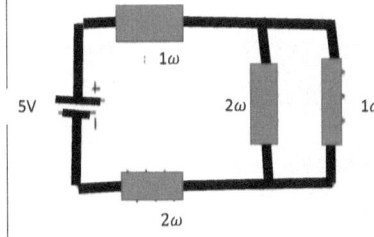

Resistor	Voltage	Current	Resistance	Power
1			1 ω	
2			2 ω	
3			1 ω	
4			2 ω	
Total				

If the Resistors were Bulbs which one will shine brightest?

_____

If the Resistors were Bulbs which one will be the dimmest?

_____

Using the following Circuit fill the table:

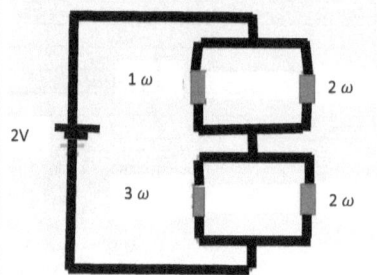

Resistor	Voltage	Current	Resistance	Power
1			1 ω	
2			2 ω	
3			3 ω	
4			2 ω	
Total				

If the Resistors were Bulbs which one will shine brightest?

_____

If the Resistors were Bulbs which one will be the dimmest?

_____

WORK:

Using the following Circuit fill the table:

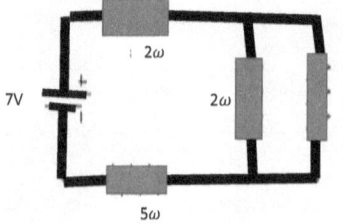

Resistor	Voltage	Current	Resistance	Power
1			2 ω	
2			2 ω	
3			9 ω	
4			5 ω	
Total				

If the Resistors were Bulbs which one will shine brightest?

_____

If the Resistors were Bulbs which one will be the dimmest?

_____

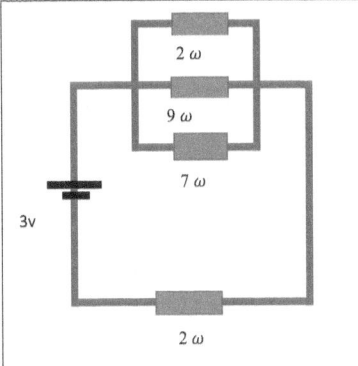

Resistor	Voltage	Current	Resistance	Power
1			2 ω	
2			9 ω	
3			7 ω	
4			2 ω	
Total				

If the Resistors were Bulbs which one will shine brightest?

_____

If the Resistors were Bulbs which one will be the dimmest?

_____

WORK:

Using the following Circuit fill the table:

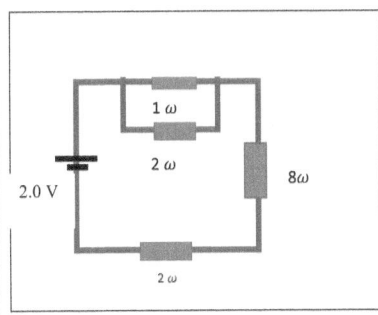

Resistor	Voltage	Current	Resistance	Power
1			1 ω	
2			2 ω	
3			8 ω	
4			2 ω	
Total				

If the Resistors were Bulbs which one will shine brightest?

_____

If the Resistors were Bulbs which one will be the dimmest?

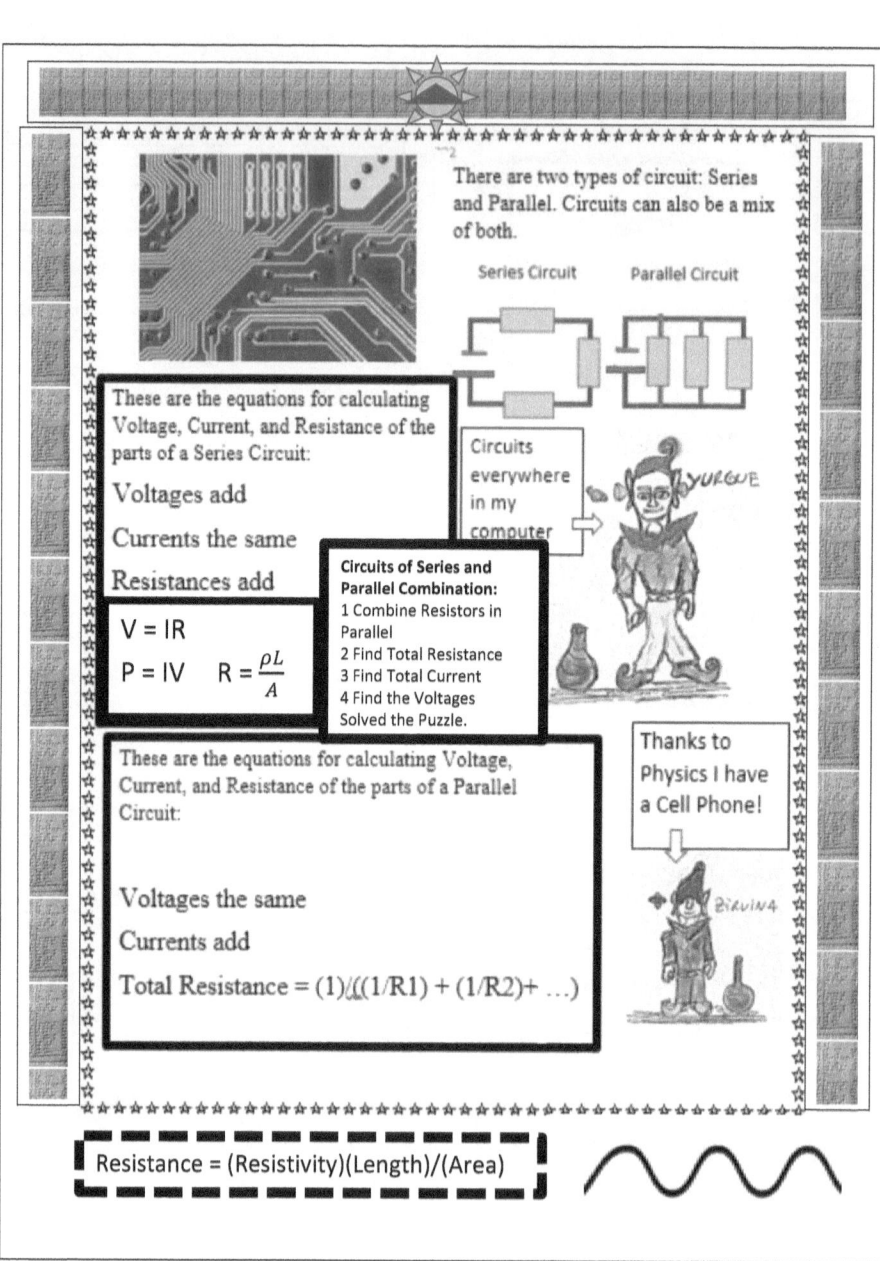

Name_____ Period_____

Circuits Review 4

Table of Resistivities:

1...Answer the following questions using the table on the right:

VVVVVVVVVVVVVVVVVVVVVVVVV

Silver: $1.59 \times 10^{-8} \, \omega m$

Copper: $1.68 \times 10^{-8} \, \omega m$

Aluminum: $2.65 \times 10^{-8} \, \omega m$

Wow! Let us venture!

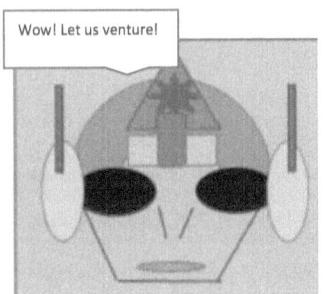

First you must find the Resistance of each resistor.

Silver
Length: 0.03m
Area: (3×10^{-11}) m^2

Silver Resistance:_____

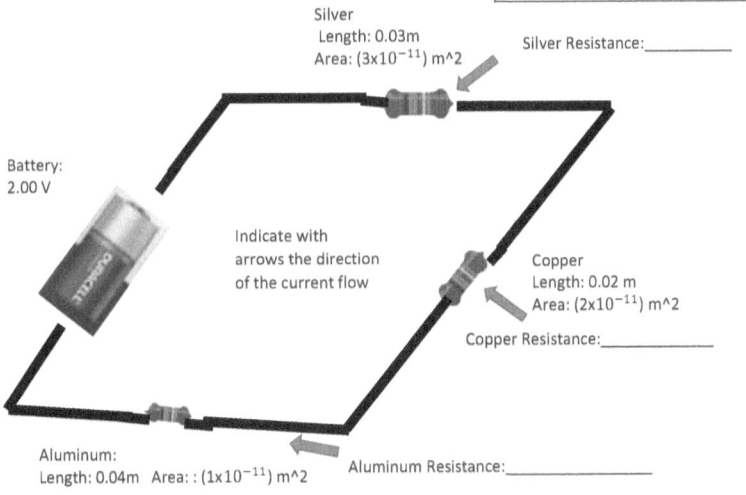

Battery:
2.00 V

Indicate with arrows the direction of the current flow

Copper
Length: 0.02 m
Area: (2×10^{-11}) m^2

Copper Resistance:_____

Aluminum:
Length: 0.04m   Area: : (1×10^{-11}) m^2

Aluminum Resistance:_____

Fill the Table below for the circuit in Page 1:

Resistor	Voltage	Current	Resistor	Power
Total				

If the Resistors were Bulbs which one will shine brightest?

_____

If the Resistors were Bulbs which one will be the dimmest?

_____

State what happens to resistance when the following changes are made to the resistor:

A….Length doubles:_____
B…Length cut to a half:_____
C….Area triples:_____
D…Area cut to a fifth:_____

What is the Total Resistance of a Parallel Circuit with the following Resistors:

10 ohms, 10 ohms, 20 ohms, 20 ohms, 50 ohms, 70 ohms, and 10 ohms:

_____ Ohms

Using the following Circuit fill the table:

Parallel Circuit

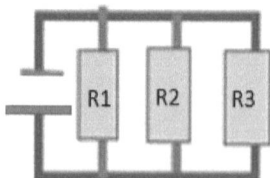

Resistor	Voltage	Current	Resistance	Power
			3.0 ω	
			4.0 ω	
			9.0 ω	
Total	9.00 V			

If the Resistors were Bulbs which one will shine brightest?

_____

If the Resistors were Bulbs which one will be the dimmest?

_____

Using the following Circuit fill the table:

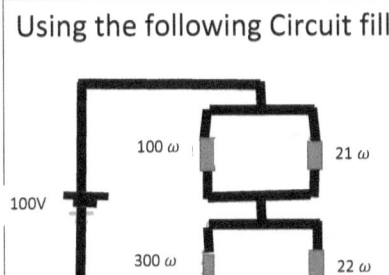

Resistor	Voltage	Current	Resistance	Power
1			100 ω	
2			21 ω	
3			300 ω	
4			22ω	
Total				

If the Resistors were Bulbs which one will shine brightest?

_____

If the Resistors were Bulbs which one will be the dimmest?

_____

WORK:

Using the following Circuit fill the table:

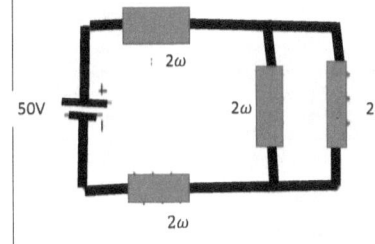

Resistor	Voltage	Current	Resistance	Power
1			2 ω	
2			2 ω	
3			2 ω	
4			2 ω	
Total				

If the Resistors were Bulbs which one will shine brightest?

_____

If the Resistors were Bulbs which one will be the dimmest?

_____

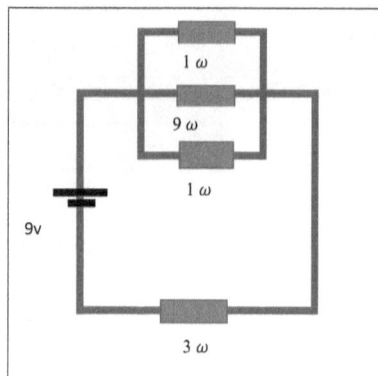

Resistor	Voltage	Current	Resistance	Power
1			1 ω	
2			9 ω	
3			1 ω	
4			3 ω	
Total				

If the Resistors were Bulbs which one will shine brightest?

_____

If the Resistors were Bulbs which one will be the dimmest?

_____

WORK:

Using the following Circuit fill the table:

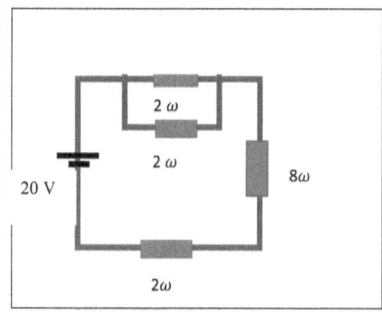

Resistor	Voltage	Current	Resistance	Power
1			2 ω	
2			2 ω	
3			8 ω	
4			2 ω	
Total				

If the Resistors were Bulbs which one will shine brightest?

_____

If the Resistors were Bulbs which one will be the dimmest?

_____

Thermal:

Four Laws of Thermodynamics:

0. Object A in thermal equilibrium with object B is also with object C; if object C is in thermal equilibrium with object B.
1. $\Delta U = Q - W$
   Change in Internal Energy is equal to the heat minus the work.
2. Entropy in the universe always increases.
3. Entropy at absolute zero is zero.

For an ideal gas:

$\frac{PV}{T} = Constant$

Heat of increasing temperature:

$Q = cm\Delta T$
$Q = heat$
$c = heat\ capacity$
$m = the\ mass\ of\ the\ material$

Latent Heat of Phase Change:

$Q = mL$
$L = Latent\ Heat$

Ideal Gas Equation:

$PV = nRT$

First Law of Thermodynamics:

$$\Delta U = Q - W$$

$\Delta U =$ Change in Internal Energy
$Q =$ Heat
$W =$ Work
$Work = \int P \cdot dv$

Cases of gas changes

Isothermal = same temperature
$\Delta U = 0$
$Q = W$
$Work = nRT \ln(\frac{V_f}{V_i})$

n = Number of Moles
R = Gas Constant
T = Temperature

Temperature is measured in Kelvins.
Kelvins = (Celsius + 273.15)

Most Efficient Energy:

1. Gas absorbs heat, undergoes isothermal expansion, and work is done by the gas. (A to B)
2. Gas expands adiabatically, temperature decreases, and work is done by the gas. (B to C)
3. Gas compresses isothermally, gives of heat, and work is done on the gas. (C to D)
4. Gas compresses adiabatically, temperature increases, and work is done on the gas. (D to A)

---- No engine is more efficient than the Carnot Engine

Work done by the gas is positive work.
Work done on the gas is negative work.

Change in Entropy in the Carnot Engine Cycle is zero

Isobaric = Same Pressure
Work = P(vf − vi)
Adiabatic = No heat flow
Q = 0, ΔU = −W
Isochoric = no volume change
W = 0
Root mean square velocity of molecules in an ideal gas:
$V = \sqrt{\frac{3RT}{Molar\ Mass}}$ from $(\frac{1}{2})mv^2 = (3/2)nRT$
Internal energy $U = \frac{3}{2}nRT$
Change in internal energy
$\Delta U = \frac{3}{2}nR\Delta T$

The Atom:

Proton, Neutron, Electron

Machines   Work = Qh − Qc

Qh = heat in the hot reservoir.
Qc = heat in the cold reservoir.
Efficiency = $\frac{Qh-Qc}{Qh} = 1 - \frac{Qc}{Qh}$
No machine is 100% efficient.
Entropy is the amount of disorder which is $\frac{Q}{T} = S$
$\Delta S = \frac{\Delta Q}{T}$ For a reversible process. All real-world processes are irreversible.
Performance of a heat pump is $\frac{Qh}{work}$
Performance of a refrigerator is $\frac{Qc}{work}$

The wave function of an electron around an atom should obey the quantum states.

The angular momentum of an electron in an atom exists within discrete nh(bar) states.

That is why $mvr = \frac{nh}{2\pi}$

Electrons around atoms can have energy levels of 1,2,3......

Heat In → Machine → Heat Out
Work ↓

Each energy level has a degree of degeneracy which are many electrons sharing the same energy although different quantum state.

For each n the orbital angular momentum known as l goes from 0 to n-1.

Each orbital angular momentum has a magnetic angular momentum known as ml that goes from −l to l.

Each ml has two possible spins which are positive $\frac{1}{2}$ or negative $\frac{1}{2}$. Positive spin stands for spin up, while negative spin stands for spin down.

There are(2l+1) numbers of mls for a given l. The degeneracy for each energy level is $2n^2$.

This degeneracy in energy levels is proven by the following:

2 spins for each (2l+1) gives:

$2(2l+1)$

Summing up all the l's from 0 to n-1 we get:

$2\sum_0^{n-1}(2l+1)$

$2(2\sum_0^{n-1}l + n)$

$2(2\frac{1}{2}n(n-1) + n)$

$2(n^2 - n + n) = 2n^2$

Particles exist in a Sea of Quantum Fluctuations

There are four forces in the universe:

Strong Force – which holds the nucleus together with gluons.
Weak Force – which breaks up the nucleus with Z and W bosons.
Electromagnetic Force – which gives charge to particles with photons.
Gravitational Force – which makes masses attract by curving space time.

There are two types of particles: Bosons and Fermions.

Fermions can't occupy the same quantum energetic state and have half integer spin.

Bosons can occupy the same place and quantum energetic state and have integer spin.

Protons, neutrons, and electrons are fermions.

Thanks to these particles being fermions the universe exists in its place, because if they were bosons nothing would have come together to form matter as we know today. Bosons can't distinguish one from the other. The Pauli Exclusion Principle in the cosmos allows matter to distinguish particles from other particles giving shape to the diversity we see around us.

Hadrons are particles made of quarks.

Mesons are hadrons composed of two quarks.

Baryons are hadrons composed of three quarks.

Protons and neutrons are baryons.

Leptons are fundamental particles made of no other particles.

Electrons are leptons, as well as the muon, the tau, and their respective neutrinos.

Tau is heavier than the muon, which is heavier than the electron.

When neutrinos travel through space, they may change from one to the other. An electron neutrino may change to a tau neutrino, and that tau neutrino back to an electron neutrino. These are called fluctuations.

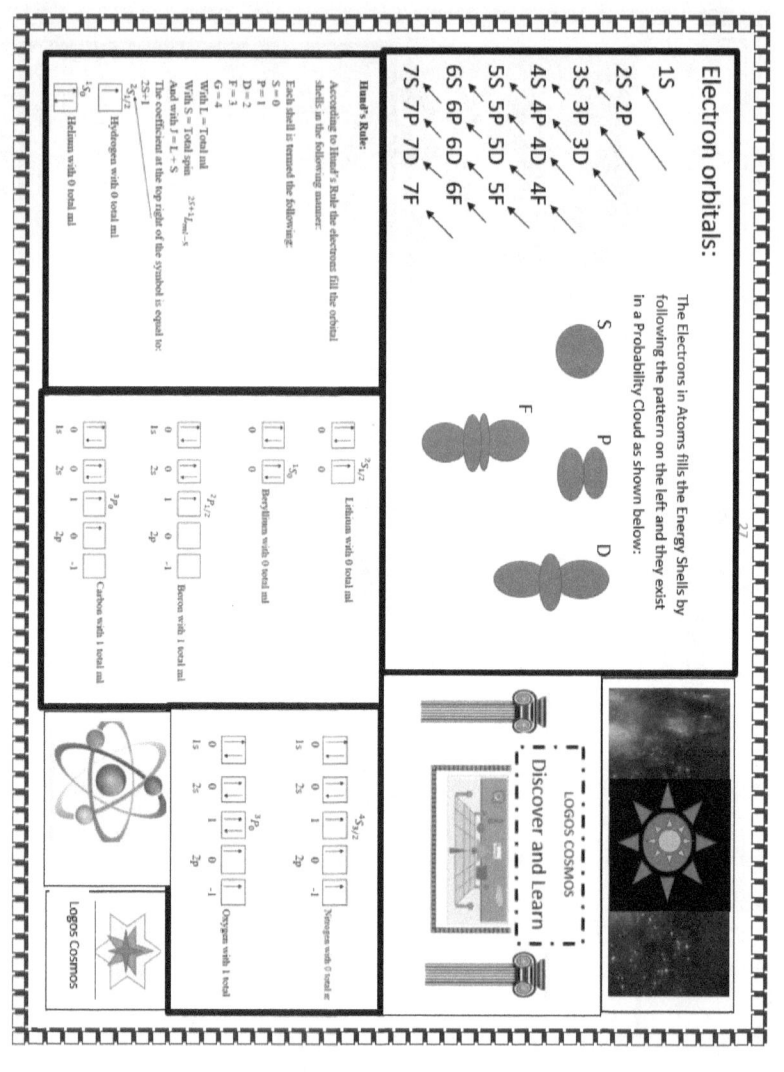

Nuclear Physics

There is a binding energy holding the particles in the nucleus together.

The binding energy can be calculated as the mass difference times c^2:

$\Delta mc^2 = c^2$ (mass of nucleus − total mass of each individual nucleon)

Nuclei of certain atoms are unstable and undergo nuclear decay.

There are three types of decay:

Alpha, beta, and gamma.

Alpha decays are emissions of helium nuclei of two protons and two neutrons.

In nuclear physics A is the number of neutrons and protons, and Z is just the number of protons.

What determines an element is its number of protons.

So when elements undergo decay, they may change to another element.

Alpha particles are positive charged because of the two protons.

Beta decays are emissions of electrons, so they are negative charged.

Gamma decays are neutral because they are photons. Photons have no charge.

Three types of Radiation in a Magnetic Field:

Magnetic Field out the page

Radiation — Path of beta

B is out the page in this square

Path of Alpha — Path of Gamma

The neutrinos in the beta decay example are included in the theory to account for the missing energy and angular momentum.

In a decay the number of nucleons, lepton number, and total charge should be preserved on both sides.

Electrons and neutrinos have lepton number of 1.

Positrons and anti-neutrinos have lepton number of -1.

To keep the net lepton number zero, an electron is followed by an antineutrino, and a positron is followed by a neutrino.

The radius of the nucleus of an atom is:

$r = (1.2 \times 10^{-15} m)(A^{\frac{1}{3}})$

Where A is the number of nucleons.

Nucleons are the Neutrons and Protons.

The activity of a material under decay is:

$Activity = -\frac{\Delta N}{\Delta t} = \frac{Change\ in\ number\ of\ nuclei}{time}$

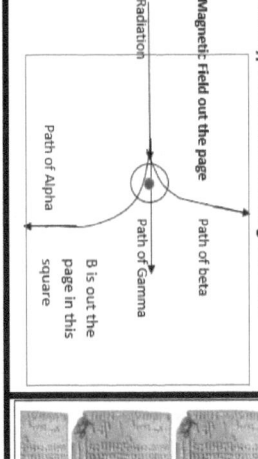

Nuclei is the nucleus.

$-\frac{\Delta N}{\Delta t} = \lambda N$, where λ is the decay constant.

Solving the differential equation, we get the following:

$N = N_0 e^{-\lambda t}$

Where N0 is the number of nuclei the material starts with, and N is the number of nuclei after some time.

The amount of time that it takes for half of the material to decay is:

$\frac{1}{2} = \frac{N}{N_0} = e^{-\lambda t}$

$\text{Ln}(\frac{1}{2}) = -\lambda t$, $-\ln(2) = -\lambda t$, $\frac{\ln(2)}{\lambda} = t$ = amount of time

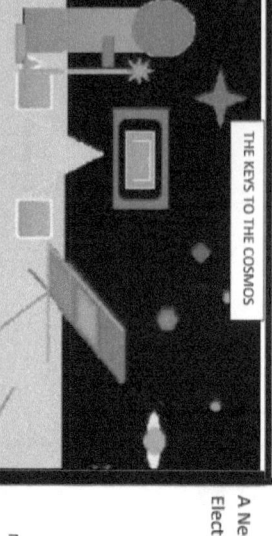

THE KEYS TO THE COSMOS

Decay of a Proton:

$P \to N + e^+ + \text{Ve}$

A Proton becomes a Neutron, a Positron, and an Electron Neutrino

Decay of a Neutron:

$N \to P + e^- + \overline{V_e}$

A Neutron becomes a Proton, an Electron, and an Electron Anti- Neutrino.

EVERYTHING IS WELL MEASURED AND CALCULATED

Nuclear Equations:

Each of the following elements are undergoing alpha decay. Fill in the equation on the right:

A.... $^{70}_{31}Ga + ^{4}_{2}He \longrightarrow$ _____

Baryon_____

Lepton_____

Charge_____

B.... $^{27}_{13}Al + ^{4}_{2}He \longrightarrow$ _____

Baryon_____

Lepton_____

Charge_____

C.... $^{31}_{15}P + ^{4}_{2}He \longrightarrow$ _____

Baryon_____

Lepton_____

Charge_____

D....$^{70}_{31}Ga \longrightarrow {}^{4}_{2}He +$ _____

Baryon _____

Lepton _____

Charge _____

E....$^{27}_{13}Al \longrightarrow {}^{4}_{2}He +$ _____

Baryon _____

Lepton _____

Charge _____

F....$^{31}_{15}P \longrightarrow {}^{4}_{2}He +$ _____

Baryon _____

Lepton _____

Charge _____

Each of the following elements are undergoing beta decay. Fill in the equation on the right:

A.... $^{70}_{31}Ga + e^-$ ⟶ _____ + _____

Baryon_____

Lepton_____

Charge_____

B.... $^{27}_{13}Al + e^-$ ⟶ _____ + _____

Baryon_____

Lepton_____

Charge_____

C.... $^{31}_{15}P + e^-$ ⟶ _____ + _____

Baryon_____

Lepton_____

Charge_____

D.... $^{70}_{31}Ga + e^+$ ⟶ _____ + _____

Baryon_____

Lepton_____

Charge_____

E.... $^{27}_{13}Al + e^+$ ⟶ _____ + _____

Baryon_____

Lepton_____

Charge_____

F.... $^{31}_{15}P + e^+$ ⟶ _____ + _____

Baryon_____

Lepton_____

Charge_____

G.... $^{70}_{31}Ga \longrightarrow e^- +$ _____ + _____
Baryon_____
Lepton_____
Charge_____

H.... $^{27}_{13}Al \longrightarrow e^- +$ _____ + _____
Baryon_____
Lepton_____
Charge_____

I.... $^{31}_{15}P \longrightarrow e^- +$ _____ + _____
Baryon_____
Lepton_____
Charge_____

K.... $^{70}_{31}Ga \longrightarrow e^+ +$ _____ + _____

Baryon_____

Lepton_____

Charge_____

L.... $^{27}_{13}Al \longrightarrow e^+ +$ _____ + _____

Baryon_____

Lepton_____

Charge_____

M.... $^{31}_{15}P \longrightarrow e^+ +$ _____ + _____

Baryon_____

Lepton_____

Charge_____

3. Show the reaction of a Proton becoming a Neutron with a Feynman Diagram for it:

Baryon_____

Lepton_____

Charge_____

Equation:

___ $\longrightarrow$ ___ + ___ + ___

4...Show the reaction of a Neutron becoming a Proton with a Feynman Diagram for it:

Baryon_____

Lepton_____

Charge_____

Equation:

___ $\longrightarrow$ ___ + ___ + ___

5. Given the fact that Protons and Neutrons are made of three quarks that can be either up or down, state their composition:

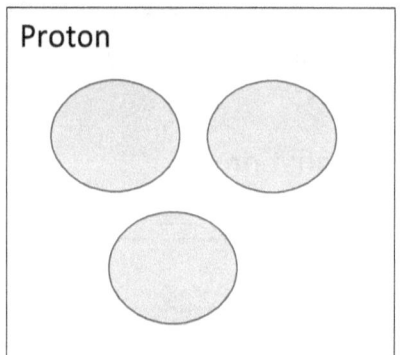

Proton

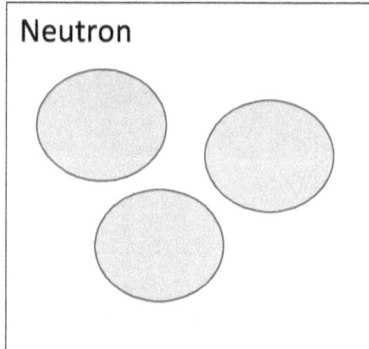

Neutron

Radiation

1…If it takes a material with 1000 atoms 3 years for half of its atoms to decay answer the following:

A…What is its T-Half Life?

B…How long will it take for the following:

Amount left	Nuclei Left	How long to decay:
1/4		
1/8		
1/16		
1/32		
1/64		

2…If it takes a material with 1000 atoms 2 years for half of its atoms to decay answer the following:

A…What is its T-Half Life?

B…How long will it take for the following:

Amount left	Nuclei Left	How long to decay:
1/4		
1/8		
1/16		
1/32		
1/64		

3…If it takes a material with 1000 atoms 6 years for half of its atoms to decay answer the following:

A…What is its T-Half Life?

> What is its Decay Constant?

B…How long will it take for the following:

Amount left	Nuclei Left	How long to decay:
50%		
40%		
35%		
30%		
20%		
10%		
5%		

4...If it takes a material with 1000 atoms 15 years for half of its atoms to decay answer the following:

A...What is its T-Half Life?

> What is its Decay Constant?

B...How long will it take for the following:

Amount left	Nuclei Left	How long to decay:
90%		
80%		
70%		
60%		
50%		
40%		
30%		

5...What are the Three Types of Radiation:
Composed of:

_____   _____

_____   _____

_____   _____

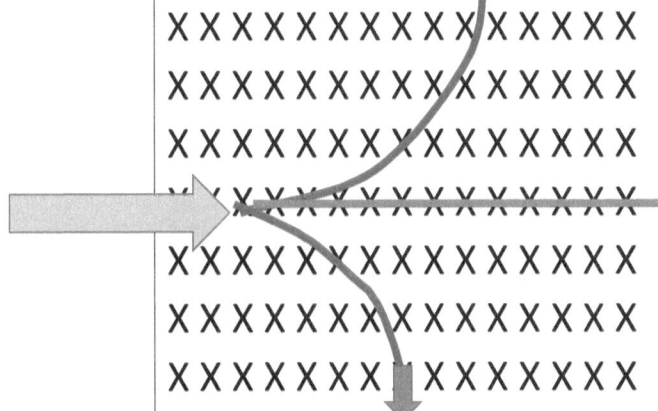

In a region of Magnetic Field Into Page what will be the path of these three Particles?

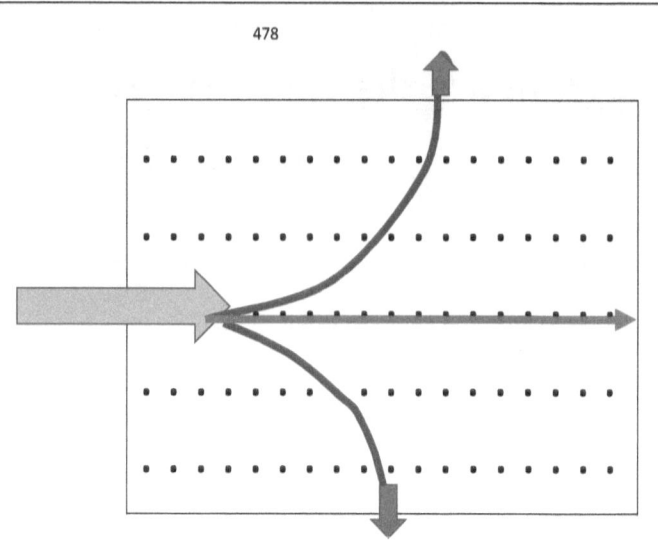

In a region of Magnetic Field Out of Page what will be the path of these three Particles?

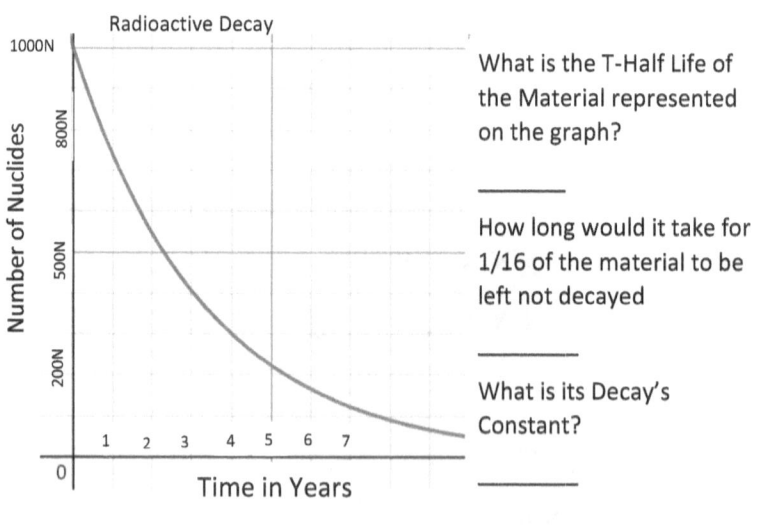

What is the T-Half Life of the Material represented on the graph?

_____

How long would it take for 1/16 of the material to be left not decayed

_____

What is its Decay's Constant?

_____

Name_____Period_____

Nuclear Equations Review 2:

Each of the following elements are undergoing alpha decay. Fill in the equation on the right:

A.... $^{32}_{16}S + ^{4}_{2}He \longrightarrow$ _____

Baryon_____

Lepton_____

Charge_____

B.... $^{35}_{17}Cl + ^{4}_{2}He \longrightarrow$ _____

Baryon_____

Lepton_____

Charge_____

C.... $^{75}_{33}As + ^{4}_{2}He \longrightarrow$ _____

Baryon_____

Lepton_____

Charge_____

D... $^{32}_{16}S \longrightarrow {}^{4}_{2}He + $ _____

Baryon_____

Lepton_____

Charge_____

E.... $^{35}_{17}Cl \longrightarrow {}^{4}_{2}He + $ _____

Baryon_____

Lepton_____

Charge_____

F.... $^{75}_{33}As \longrightarrow {}^{4}_{2}He + $ _____

Baryon_____

Lepton_____

Charge_____

Each of the following elements are undergoing beta decay. Fill in the equation on the right:

A.... $^{32}_{16}S + e^-$ ⟶ _____ + _____

Baryon_____

Lepton_____

Charge_____

B.... $^{35}_{17}Cl + e^-$ ⟶ _____ + _____

Baryon_____

Lepton_____

Charge_____

C.... $^{75}_{33}As + e^-$ ⟶ _____ + _____

Baryon_____

Lepton_____

Charge_____

D....$^{32}_{16}S + e^+$ ⟶ _____ + _____

Baryon_____

Lepton_____

Charge_____

E....$^{35}_{17}Cl + e^+$ ⟶ _____ + _____

Baryon_____

Lepton_____

Charge_____

F....$^{75}_{33}As + e^+$ ⟶ _____ + _____

Baryon_____

Lepton_____

Charge_____

G.... $^{32}_{16}S \longrightarrow e^- +$ _____ + _____

Baryon_____

Lepton_____

Charge_____

H.... $^{35}_{17}Cl \longrightarrow e^- +$ _____ + _____

Baryon_____

Lepton_____

Charge_____

I.... $^{75}_{33}As \longrightarrow e^- +$ _____ + _____

Baryon_____

Lepton_____

Charge_____

K.... $^{32}_{16}S \longrightarrow e^+ +$ _____ + _____

Baryon _____

Lepton _____

Charge _____

L.... $^{35}_{17}Cl \longrightarrow e^+ +$ _____ + _____

Baryon _____

Lepton _____

Charge _____

M.... $^{75}_{33}As \longrightarrow e^+ +$ _____ + _____

Baryon _____

Lepton _____

Charge _____

Radiation Worksheet 2

1…If it takes a material with 2000 atoms 4 years for half of its atoms to decay answer the following:

A…What is its T-Half Life?

B…How long will it take for the following:

Amount left	Nuclei Left	How long to decay:
1/4		
1/8		
1/16		
1/32		
1/64		

2…If it takes a material with 2000 atoms 3 years for half of its atoms to decay answer the following:

A…What is its T-Half Life?

B…How long will it take for the following:

Amount left	Nuclei Left	How long to decay:
1/4		
1/8		
1/16		
1/32		
1/64		

Energy Levels in an Atom:

N = 4   -0.85eV

N = 3   -1.51 eV

N = 2   -3.4 eV

N = 1   -13.6 eV

1....

A...An Electron moves from N=1 to N=4. Was Energy Absorbed or Emitted?

B...What is that Energy in Joules?

C...What is the Frequency of that Photon?

D...What is the Wavelength of that Photon?

2....

A...An Electron moves from N=2 to N=4. Was Energy Absorbed or Emitted?

B...What is that Energy in Joules?

C...What is the Frequency of that Photon?

D...What is the Wavelength of that Photon?

3....

A...An Electron moves from N=3 to N=4. Was Energy Absorbed or Emitted?

B...What is that Energy in Joules?

C...What is the Frequency of that Photon?

D...What is the Wavelength of that Photon?

4....

A...An Electron moves from N=4 to N=1. Was Energy Absorbed or Emitted?

B...What is that Energy in Joules?

C...What is the Frequency of that Photon?

D...What is the Wavelength of that Photon?

5....

A...An Electron moves from N=4 to N=2. Was Energy Absorbed or Emitted?

B...What is that Energy in Joules?

C...What is the Frequency of that Photon?

D...What is the Wavelength of that Photon?

6....

A...An Electron moves from N=4 to N=3. Was Energy Absorbed or Emitted?

B...What is that Energy in Joules?

C...What is the Frequency of that Photon?

D...What is the Wavelength of that Photon?

7....

A...An Electron moves from N=3 to N=2. Was Energy Absorbed or Emitted?

B...What is that Energy in Joules?

C...What is the Frequency of that Photon?

D...What is the Wavelength of that Photon?

8....

A...An Electron moves from N=1 to N=3. Was Energy Absorbed or Emitted?

B...What is that Energy in Joules?

C...What is the Frequency of that Photon?

D...What is the Wavelength of that Photon?

9....

A...An Electron moves from N=2 to N=4. Was Energy Absorbed or Emitted?

B...What is that Energy in Joules?

C...What is the Frequency of that Photon?

D...What is the Wavelength of that Photon?

Radioactive Decay

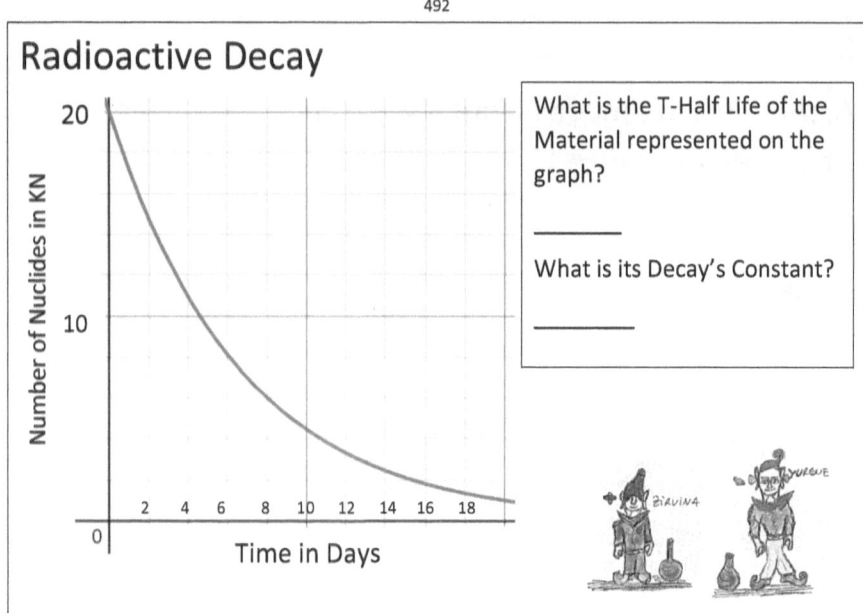

What is the T-Half Life of the Material represented on the graph?

_____

What is its Decay's Constant?

_____

How long will it take for the following:

Amount left	Nuclei Left	How long to decay:
1/4		
1/8		
1/16		
1/32		
1/64		

Radioactive Decay

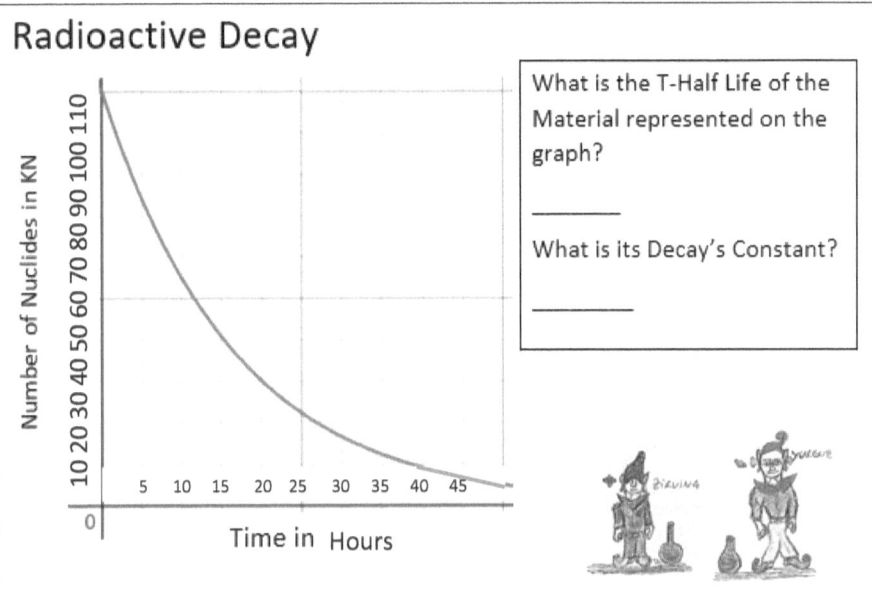

What is the T-Half Life of the Material represented on the graph?

_____

What is its Decay's Constant?

_____

How long will it take for the following:

Amount left	Nuclei Left	How long to decay:
1/4		
1/8		
1/16		
1/32		
1/64		

Extra Problems:

Name_____ Period_____

Circuits Test

Table of Resistivities:

1...Answer the following questions using the table on the right:

VVVVVVVVVVVVVVVVVVVVVVVVVV

Silver: $1.59 \times 10^{-8} \, \omega m$

Copper: $1.68 \times 10^{-8} \, \omega m$

Aluminum: $2.65 \times 10^{-8} \, \omega m$

Wow! Let us venture!

All answers in Test must be in 3 sig figs

First you must find the Resistance of each resistor.

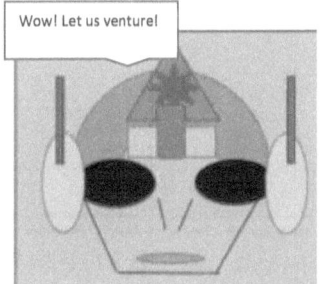

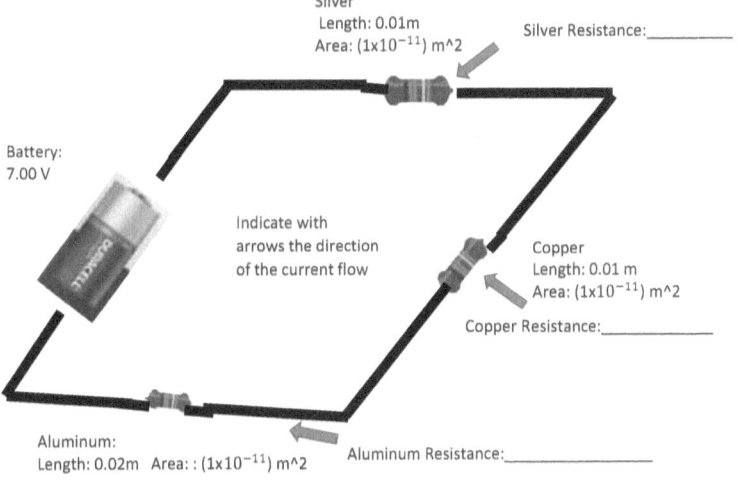

Silver
Length: 0.01m
Area: (1×10^{-11}) m^2

Silver Resistance:_____

Battery:
7.00 V

Indicate with arrows the direction of the current flow

Copper
Length: 0.01 m
Area: (1×10^{-11}) m^2

Copper Resistance:_____

Aluminum:
Length: 0.02m   Area: : (1×10^{-11}) m^2

Aluminum Resistance:_____

Fill the Table below for the circuit in Page 1:

Resistor	Voltage	Current	Resistor	Power
Total				

If the Resistors were Bulbs which one will shine brightest?

_____

If the Resistors were Bulbs which one will be the dimmest?

_____

State what happens to resistance when the following changes are made to the resistor:

A....Length triples:_____
B...Length cut to a third:_____
C....Area triples:_____
D...Area cut to a third:_____

What is the Total Resistance of a Parallel Circuit with the following Resistors:

1 ohms, 1 ohms, 2 ohms, 2 ohms, 5 ohms, 7 ohms, and 1 ohms:

_____ Ohms

Using the following Circuit fill the table:

Parallel Circuit

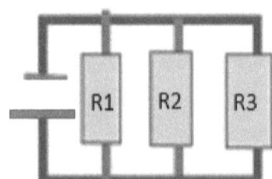

Resistor	Voltage	Current	Resistance	Power
			1.0ω	
			2.0ω	
			3.0ω	
Total	8.00 V			

If the Resistors were Bulbs which one will shine brightest?

_____

If the Resistors were Bulbs which one will be the dimmest?

_____

Using the following Circuit fill the table:

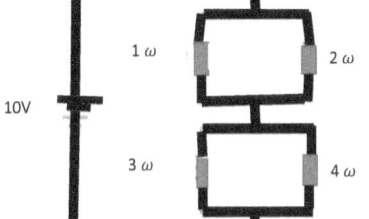

10V

Resistor	Voltage	Current	Resistance	Power
1			1 ω	
2			2 ω	
3			3 ω	
4			4ω	
Total				

If the Resistors were Bulbs which one will shine brightest?

_____

If the Resistors were Bulbs which one will be the dimmest?

_____

WORK:

Using the following Circuit fill the table:

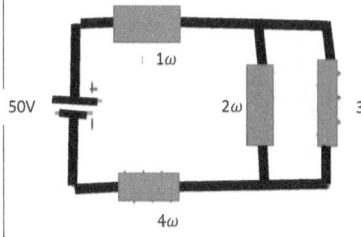

50V

Resistor	Voltage	Current	Resistance	Power
1			1 ω	
2			2 ω	
3			3ω	
4			4ω	
Total				

If the Resistors were Bulbs which one will shine brightest?

_____

If the Resistors were Bulbs which one will be the dimmest?

_____

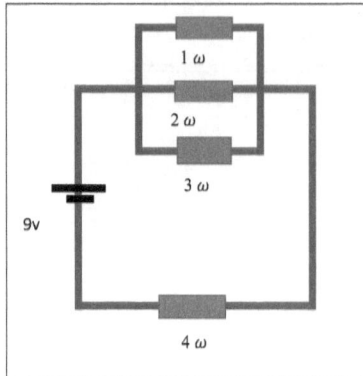

Resistor	Voltage	Current	Resistance	Power
1			1 ω	
2			2 ω	
3			3 ω	
4			4 ω	
Total				

If the Resistors were Bulbs which one will shine brightest?

_____

If the Resistors were Bulbs which one will be the dimmest?

_____

WORK:

Using the following Circuit fill the table:

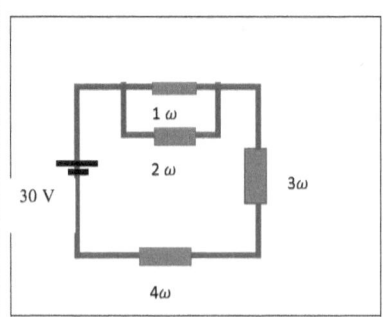

Resistor	Voltage	Current	Resistance	Power
1			1ω	
2			2 ω	
3			3 ω	
4			4 ω	
Total				

If the Resistors were Bulbs which one will shine brightest?

_____

If the Resistors were Bulbs which one will be the dimmest?

Name _____ Period _____

Right Hand Rule Review 2

1… What is the Electric Field at a Distance of 2.9mm from a Particle with charge 1.8C?

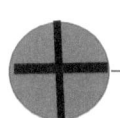 ——————————— 2.9 mm

2… Does the Field in Problem 1 point away or towards the Particle? _____

3… If now a Particle with 4.6cC of charge is placed at that same location from the 1.8C Particle, what is the Magnitude of the Force between them?

4...Is the Force Attractive of Repulsive?

_____

5...A loop is placed in a Magnetic Field with current in the direction shown with the arrows.

The Physics of Electric Motors

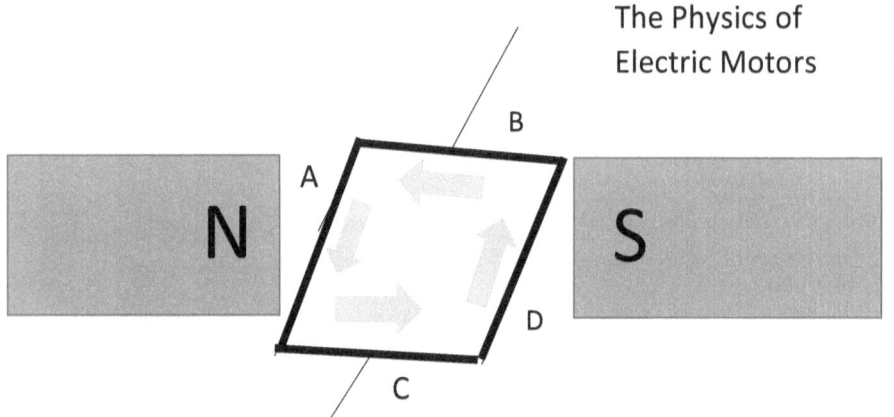

This set up will cause the loop to spin. State the direction of the Force on the side A and Side D of the Loop:

Side A_____

Side D _____

To keep the loop rotating with full revolutions, the directions of the Currents are inverted many times.

6…A loop is placed in a Magnetic Field with current in the direction shown with the arrows.

The Physics of Electric Motors

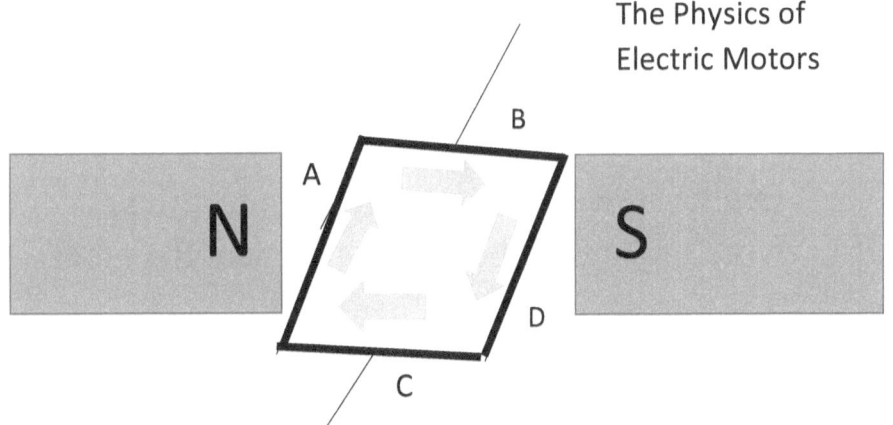

This set up will cause the loop to spin. State the direction of the Force on the side A and Side D of the Loop:

Side A_____

Side D _____

7…You now insert a magnet with its South side into a loop of wire.

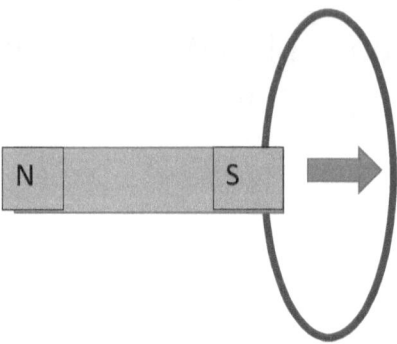

Draw arrows around the loop indicating which way will the Current flow in the wire?

The universe is composed of many loops of Electric and Magnetic Fields.

8...Draw arrows around the loop indicating which way will the current flow in the wire if now you are pulling the magnet out of the loop with its N side facing the loop:

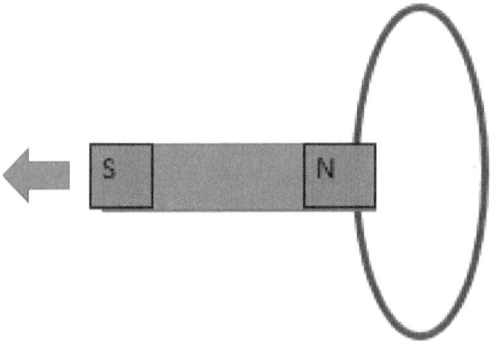

Write which of the Maxwell's Equations describes this phenomenon:

_____

9...Draw arrows around the loop indicating which way will the current flow in the wire if now you are pulling the magnet out of the loop with its S side facing the loop:

Write which of the Maxwell's Equations describes this phenomenon:

_____

10…You now insert a magnet with its North side into a loop of wire.

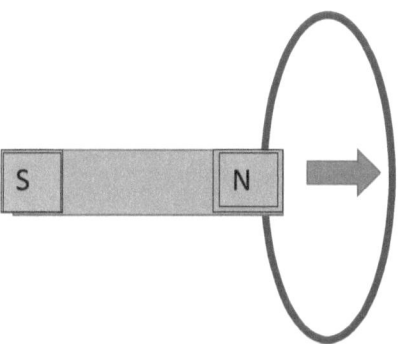

Draw arrows around the loop indicating which way will the Current flow in the wire?

The universe is composed of many loops of Electric and Magnetic Fields.

11...A loop is placed in a Magnetic Field with current in the direction shown with the arrows.

The Physics of Electric Motors

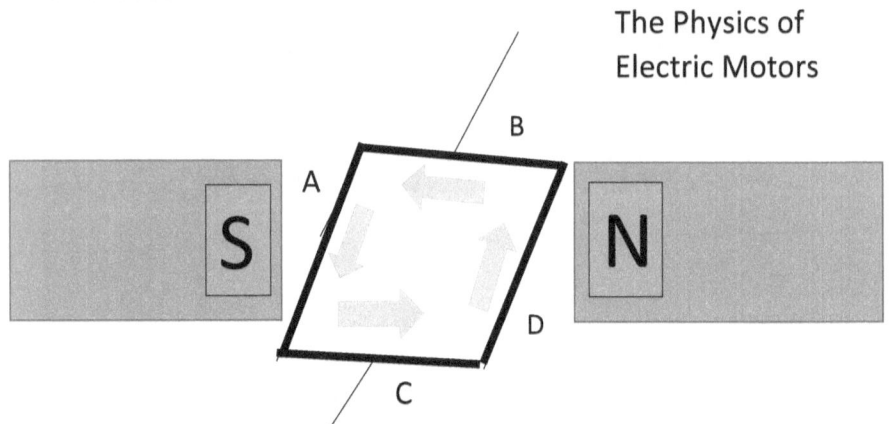

This set up will cause the loop to spin. State the direction of the Force on the side A and Side D of the Loop:

Side A_____

Side D _____

12…A loop is placed in a Magnetic Field with current in the direction shown with the arrows.

The Physics of Electric Motors

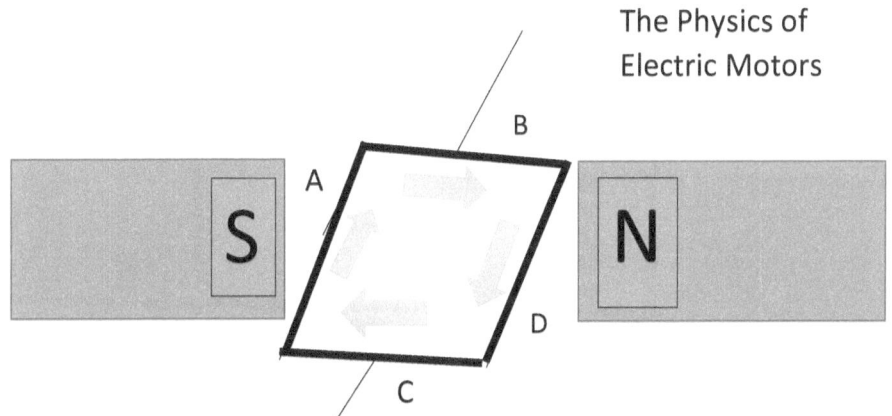

This set up will cause the loop to spin. State the direction of the Force on the side A and Side D of the Loop:

Side A_____

Side D _____

13...Explain which pair of wires will there be a force of attraction between the wires?

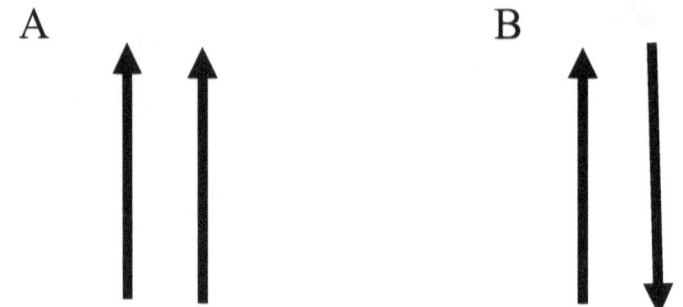

14...Use dots and Xs to describe the Magnetic Field around the following wires:

15...Draw an arrow for the direction of the Magnetic Field produced in these Solenoids (Coils) with current in the direction shown:

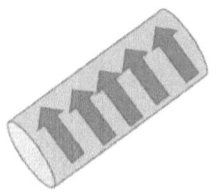

16... What is the Magnetic Field from a wire carrying 0.8A of Current at a distance 2.6 mm away?

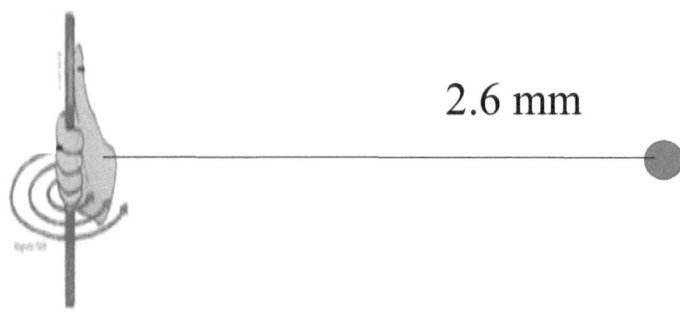

2.6 mm

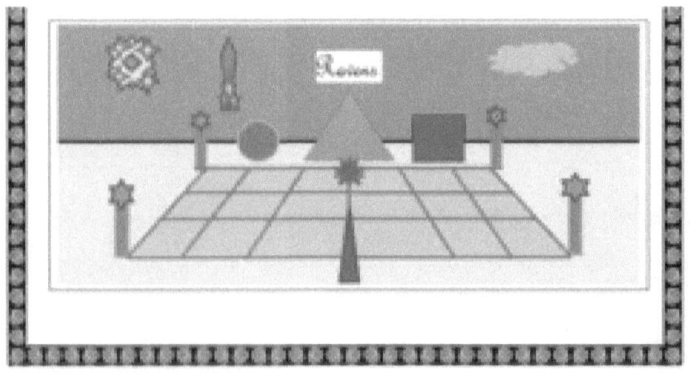

17…Explain what each equation represents:

$\nabla \cdot \mathbf{E} = \dfrac{\rho}{\varepsilon_0}$

$\nabla \cdot \mathbf{B} = 0$

$\nabla \times \mathbf{E} = -\dfrac{\partial \mathbf{B}}{\partial t}$

$\nabla \times \mathbf{B} = \mu_0 \mathbf{j} + \dfrac{1}{c^2}\dfrac{\partial \mathbf{E}}{\partial t}$

ZiRViN4

Light is an Electromagnetic Wave that spins and twists as it travels through space. Many things spin in space. Stars, Planets, Particles, and Galaxies. Spinning is the way the cosmos dance in this Universal Flow.

512

18...Draw a Force Vector for the Wires carrying Current in the following directions

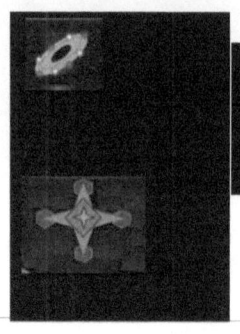

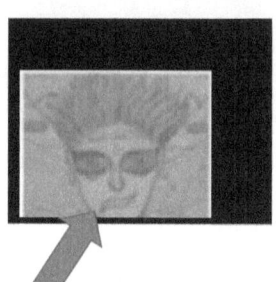

Space is filled with Electric and Magnetic Fields:

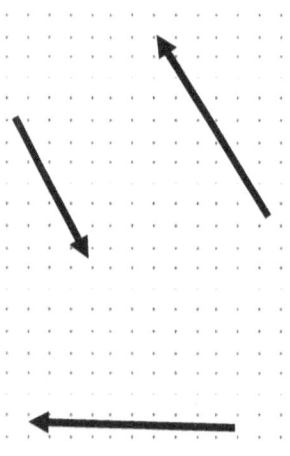

x x x x x x x x x x x x x x x x x x
x x x x x x x x x x x x x x x x x x
x x x x x x x x x x x x x x x x x x
x x x x x x x x x x x x x x x x x x
x x x x x x x x x x x x x x x x x x
x x x x x x x x x x x x x x x x x x
x x x x x x x x x x x x x x x x x x
x x x x x x x x x x x x x x x x x x
x x x x x x x x x x x x x x x x x x
x x x x x x x x x x x x x x x x x x
x x x x x x x x x x x x x x x x x x
x x x x x x x x x x x x x x x x x x

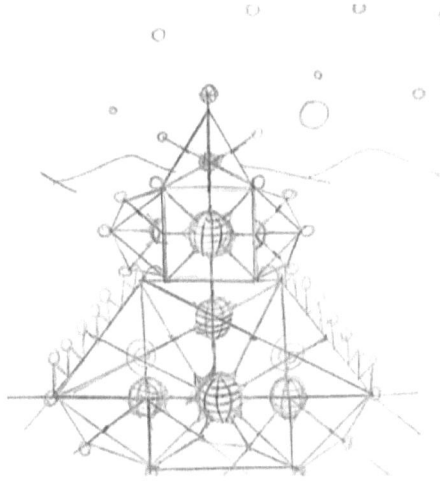

Mathematics and Geometry are in the Foundation of all the Laws and Constants found throughout the universe.

19...A Particle moves in a circle shown due to a Magnetic Field as demonstrated. Using the right-hand rule and taking note that the Force is towards the center of the circle, is the Particle positive of negative?

_____

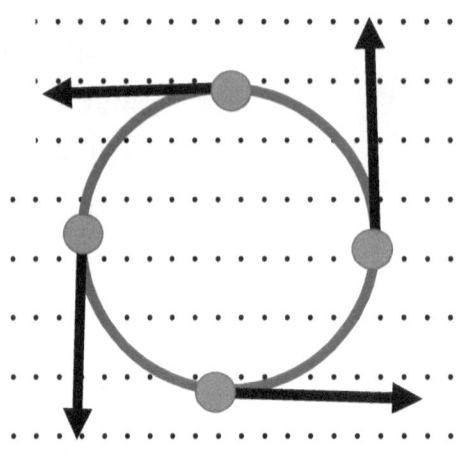

20...A Particle moves in a circle shown due to a Magnetic Field as demonstrated. Using the right-hand rule and taking note that the Force is towards the center of the circle, is the Particle positive of negative?

_____

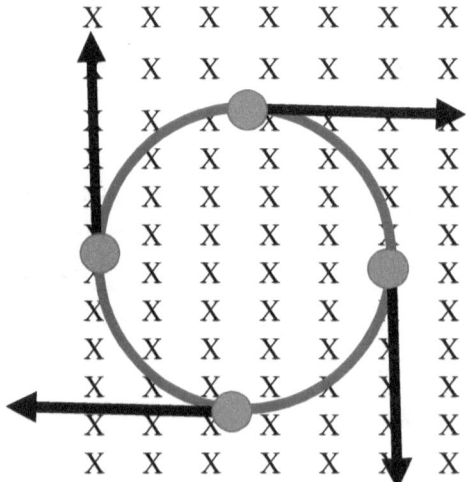

21...A Particle moves in a circle shown due to a Magnetic Field as demonstrated. Using the right-hand rule and taking note that the Force is towards the center of the circle, is the Particle positive of negative?

_____

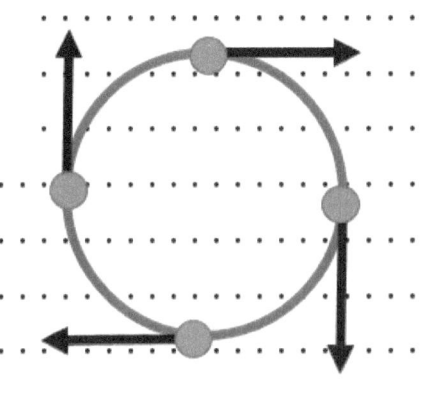

22. ..A Particle moves in a circle shown due to a Magnetic Field as demonstrated. Using the right-hand rule and taking note that the Force is towards the center of the circle, is the Particle positive of negative?

_____

21…Label the Colors of Light from Most Energetic to Least Energetic:

Most Energetic: _____

_____

_____

_____

_____

_____

Least Energetic: _____

Formula Sheet:

$$E = \frac{kQ}{r^2} \qquad F = Eq \qquad B = \frac{\mu_0 I}{2\pi r}$$

$$F = QVB \qquad F = ILB \qquad \mu_0 = 4\pi \times 10^{-7}$$

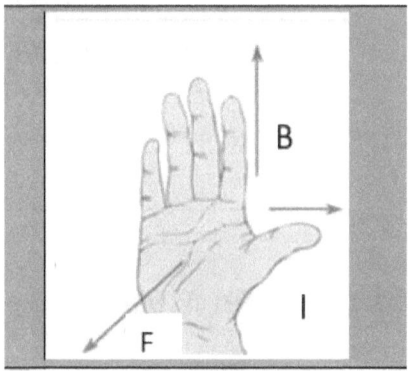

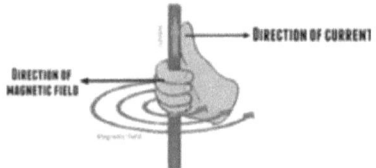

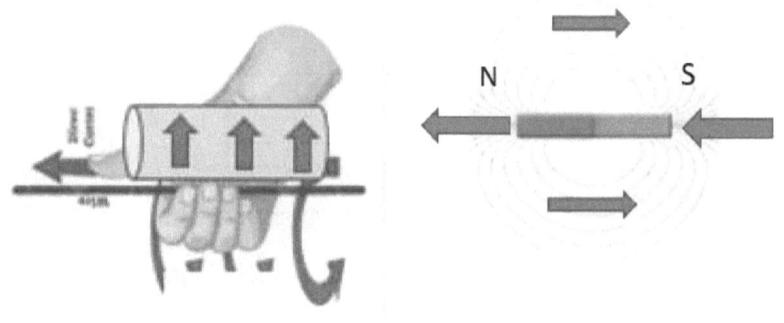

Name_____Period___

Nuclear Physics Test

1...Fill in the equation on the right:

A....$^{23}_{11}Na + ^{4}_{2}He \longrightarrow$ _____

Baryon_____

Lepton_____

Charge_____

2...$^{23}_{11}Na \longrightarrow ^{4}_{2}He +$ _____

Baryon_____

Lepton_____

Charge_____

3...$^{23}_{11}Na + e^{-} \longrightarrow$ _____ + _____

Baryon_____

Lepton_____

Charge_____

4.... $^{23}_{11}Na + e^+ \longrightarrow$ _____ + _____

Baryon_____

Lepton_____

Charge_____

5.... $^{23}_{11}Na \longrightarrow e^- +$ _____ + _____

Baryon_____

Lepton_____

Charge_____

6.... $^{23}_{11}Na \longrightarrow e^+ +$ _____ + _____

Baryon_____

Lepton_____

Charge_____

7...What are the Three Types of Radiation:

Composed of:

_____   _____

_____   _____

_____   _____

```
X X X X X X X X X X X X X X X
X X X X X X X X X X X X X X X
X X X X X X X X X X X X X X X
X X X X X X X X X X X X X X X
X X X X X X X X X X X X X X X
X X X X X X X X X X X X X X X
X X X X X X X X X X X X X X X
```

In a region of Magnetic Field Into Page what will be the path of these three Particles?

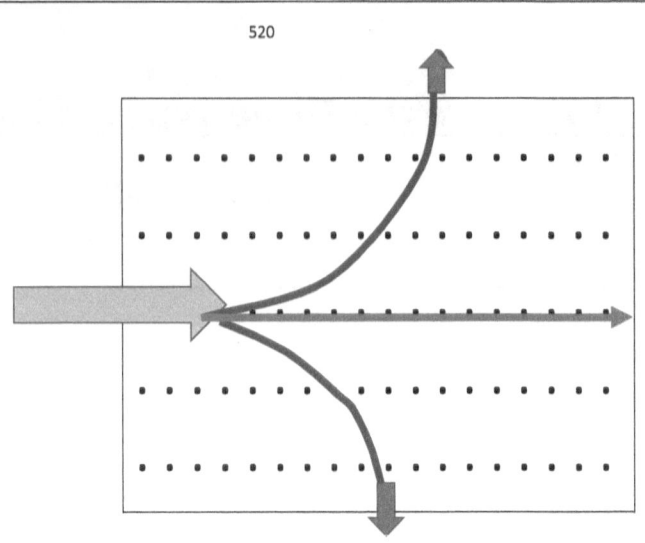

In a region of Magnetic Field Out of Page what will be the path of these three Particles?

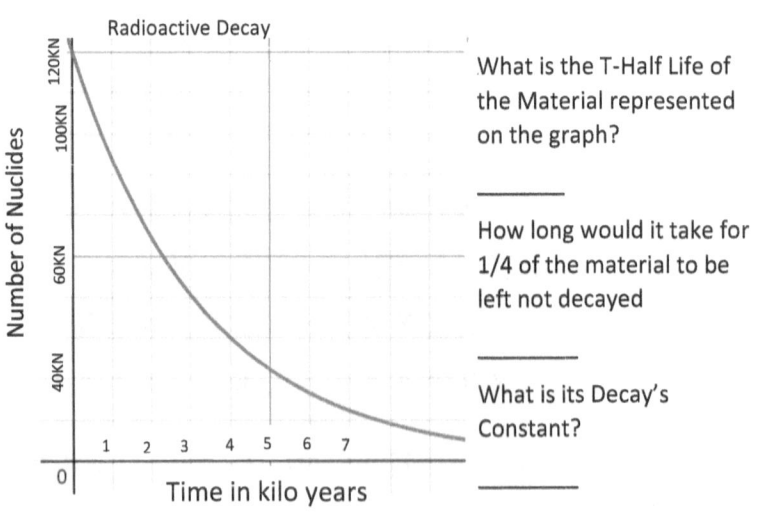

What is the T-Half Life of the Material represented on the graph?

_____

How long would it take for 1/4 of the material to be left not decayed

_____

What is its Decay's Constant?

_____

8…How long will it take for the following:

Amount left	Nuclei Left	How long to decay:
1/4		
1/8		
1/16		
1/32		
1/64		

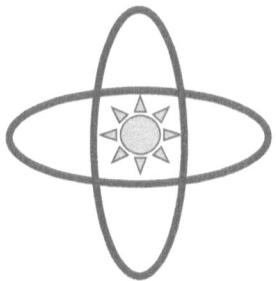

Energy Levels in an Atom:

N = 4 -0.85eV
N = 3 -1.51 eV

N = 2 -3.4 eV

N = 1 -13.6 eV

9....

A...An Electron moves from N=1 to N=4. Was Energy Absorbed or Emitted?

B...What is that Energy in Joules?

C...What is the Frequency of that Photon?

D...What is the Wavelength of that Photon?

10....

A...An Electron moves from N=3 to N=2. Was Energy Absorbed or Emitted?

B...What is that Energy in Joules?

C...What is the Frequency of that Photon?

D...What is the Wavelength of that Photon?

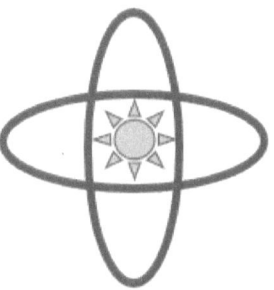

Radioactive Decay

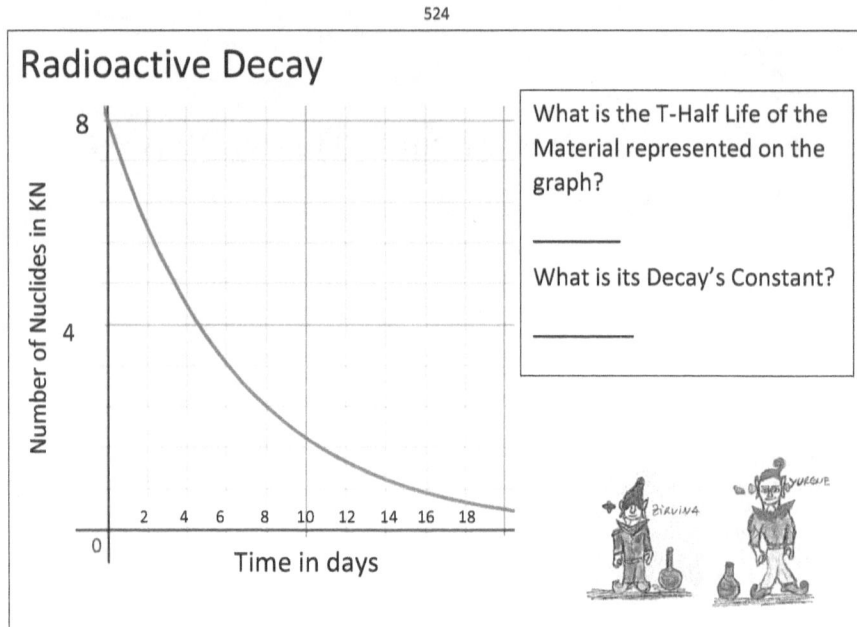

What is the T-Half Life of the Material represented on the graph?

_____

What is its Decay's Constant?

_____

How long will it take for the following:

Amount left	Nuclei Left	How long to decay:
1/4		
1/8		
1/16		
1/32		
1/64		

Name _____ Period _____

Right Hand Rule Test

1…What is the Electric Field at a Distance of 3.9cm from a Particle with charge -0.8cC?

2…Does the Field in Problem 1 point away or towards the Particle? _____

3…If now a Particle with 4.6C of charge is placed at that same location from the -0.8cC Particle, what is the Magnitude of the Force between them?

4...Is the Force Attractive of Repulsive?

_____

5...A loop is placed in a Magnetic Field with current in the direction shown with the arrows.

The Physics of Electric Motors

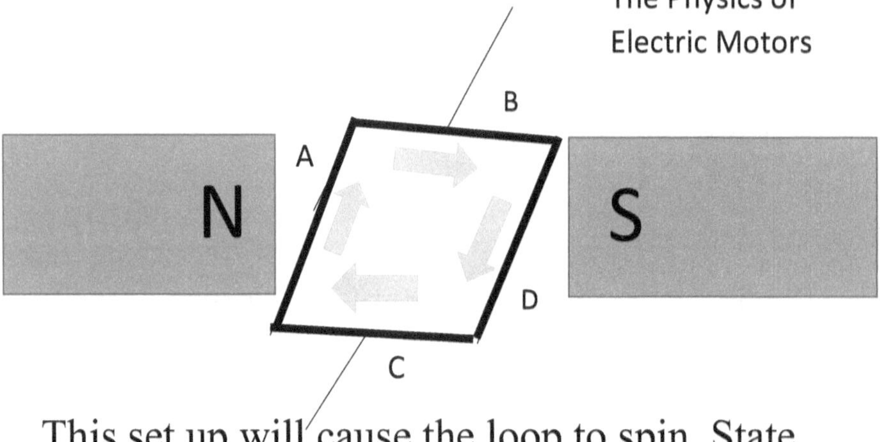

This set up will cause the loop to spin. State the direction of the Force on the side A and Side D of the Loop:

Side A_____

Side D _____

To keep the loop rotating with full revolutions, the directions of the Currents are inverted many times.

6…A loop is placed in a Magnetic Field with current in the direction shown with the arrows.

The Physics of Electric Motors

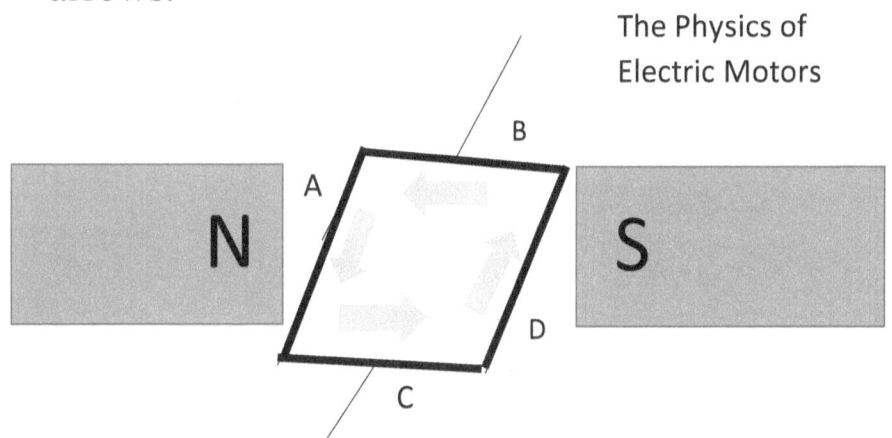

This set up will cause the loop to spin. State the direction of the Force on the side A and Side D of the Loop:

Side A_____

Side D _____

7…You now insert a magnet with its North side into a loop of wire.

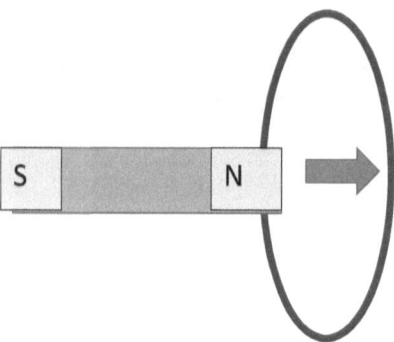

Draw arrows around the loop indicating which way will the Current flow in the wire?

The universe is composed of many loops of Electric and Magnetic Fields.

8…Draw arrows around the loop indicating which way will the current flow in the wire if now you are pulling the magnet out of the loop with its S side facing the loop:

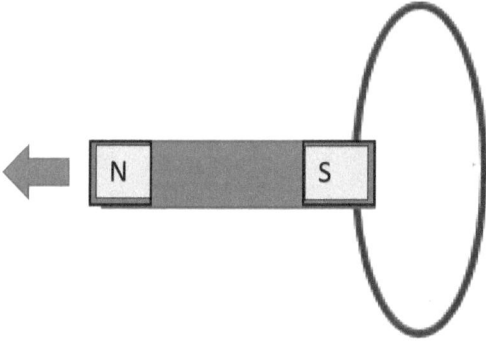

Write which of the Maxwell's Equations describes this phenomenon:

_____

9...Draw arrows around the loop indicating which way will the current flow in the wire if now you are pulling the magnet out of the loop with its N side facing the loop:

Write which of the Maxwell's Equations describes this phenomenon:

_____

10…You now insert a magnet with its South side into a loop of wire.

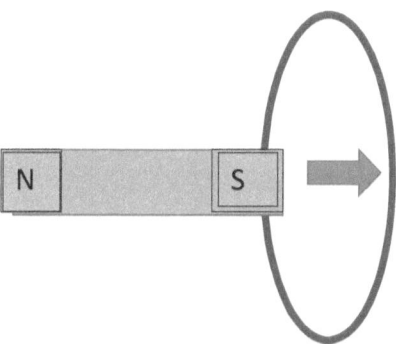

Draw arrows around the loop indicating which way will the Current flow in the wire?

The universe is composed of many loops of Electric and Magnetic Fields.

11...A loop is placed in a Magnetic Field with current in the direction shown with the arrows.

The Physics of Electric Motors

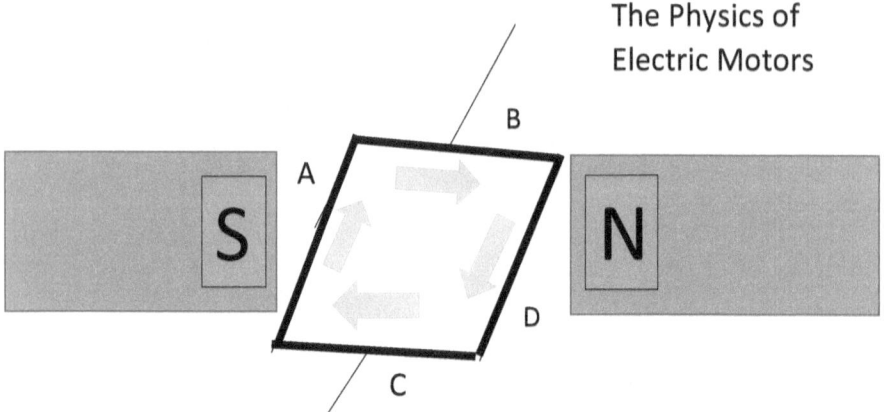

This set up will cause the loop to spin. State the direction of the Force on the side A and Side D of the Loop:

Side A_____

Side D _____

12…A loop is placed in a Magnetic Field with current in the direction shown with the arrows.

The Physics of Electric Motors

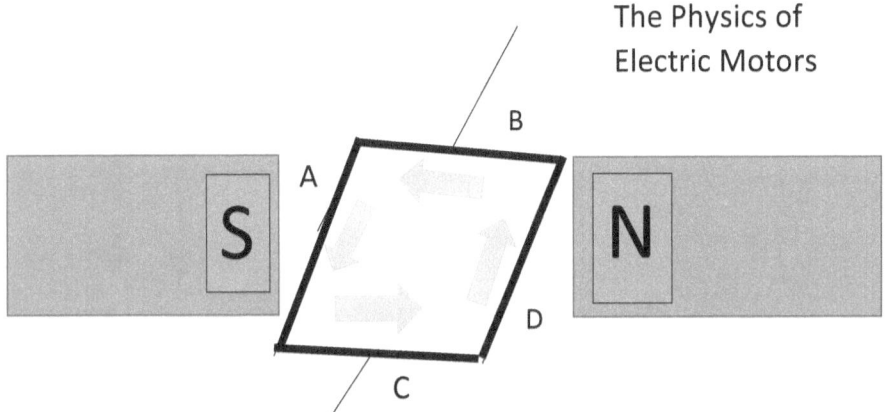

This set up will cause the loop to spin. State the direction of the Force on the side A and Side D of the Loop:

Side A_____

Side D _____

13...Explain which pair of wires will there be a force of attraction between the wires?

A   B

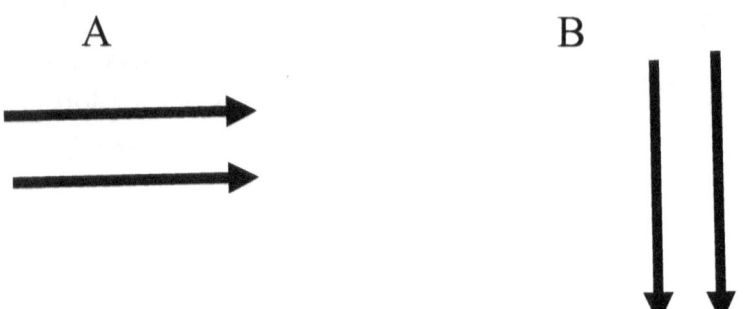

14...Use dots and Xs to describe the Magnetic Field around the following wires:

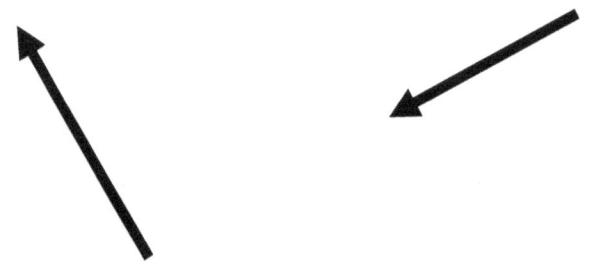

15...Draw an arrow for the direction of the Magnetic Field produced in these Solenoids (Coils) with current in the direction shown:

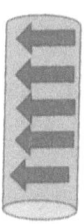

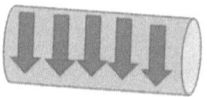

16…What is the Magnetic Field from a wire carrying 2.8A of Current at a distance 1.6 cm away?

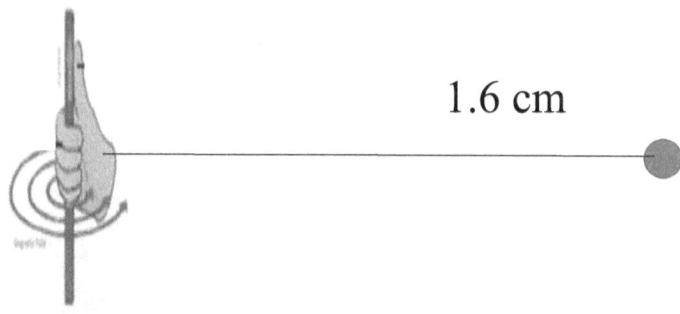

1.6 cm

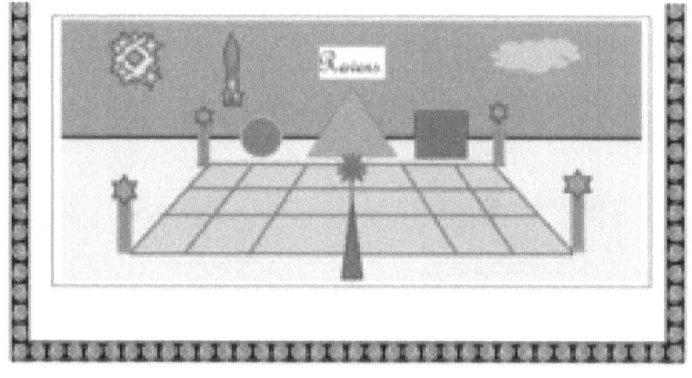

17...Explain what each equation represents:

$$\nabla \cdot \mathbf{E} = \frac{\rho}{\varepsilon_0}$$

$$\nabla \cdot \mathbf{B} = 0$$

$$\nabla \times \mathbf{E} = -\frac{\partial \mathbf{B}}{\partial t}$$

$$\nabla \times \mathbf{B} = \mu_0 \mathbf{j} + \frac{1}{c^2} \frac{\partial \mathbf{E}}{\partial t}$$

ZIRVINA

Light is an Electromagnetic Wave that spins and twists as it travels through space. Many things spin in space. Stars, Planets, Particles, and Galaxies. Spinning is the way the cosmos dance in this Universal Flow.

18...Draw a Force Vector for the Wires carrying Current in the following directions

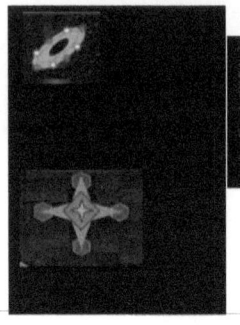

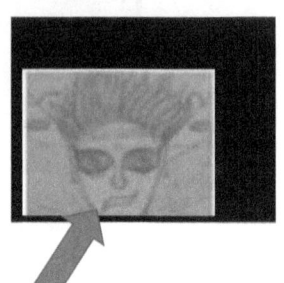

Space is filled with Electric and Magnetic Fields:

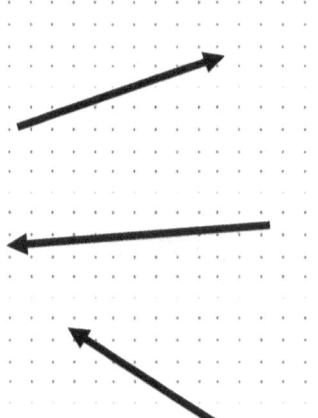

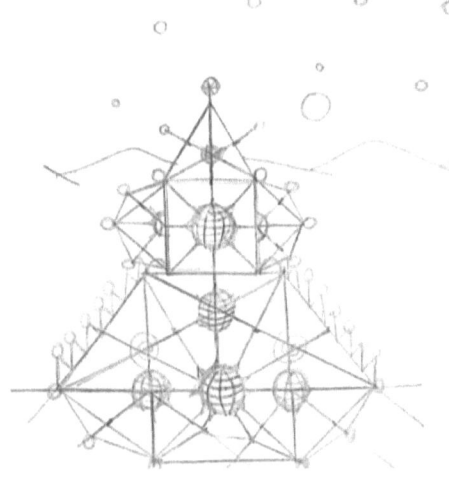

Mathematics and Geometry are in the Foundation of all the Laws and Constants found throughout the universe.

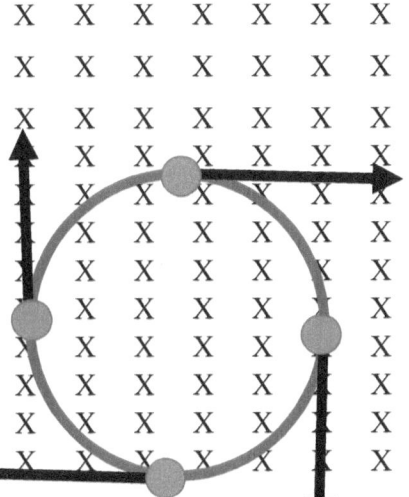

19...A Particle moves in a circle shown due to a Magnetic Field as demonstrated. Using the right-hand rule and taking note that the Force is towards the center of the circle, is the Particle positive of negative?

_____

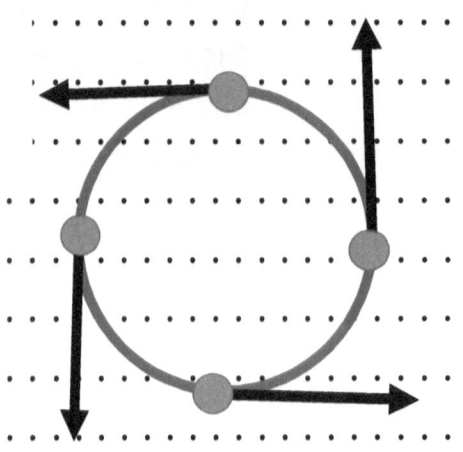

20...A Particle moves in a circle shown due to a Magnetic Field as demonstrated. Using the right-hand rule and taking note that the Force is towards the center of the circle, is the Particle positive of negative?

_____

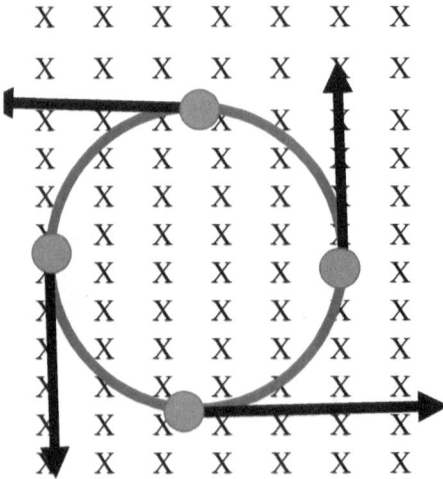

21...A Particle moves in a circle shown due to a Magnetic Field as demonstrated. Using the right-hand rule and taking note that the Force is towards the center of the circle, is the Particle positive of negative?

_____

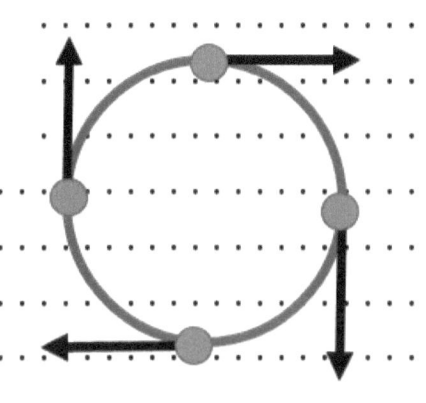

22. ..A Particle moves in a circle shown due to a Magnetic Field as demonstrated. Using the right-hand rule and taking note that the Force is towards the center of the circle, is the Particle positive of negative?

_____

21…Label the Colors of Light from Most Energetic to Least Energetic:

Most Energetic: _____

_____

_____

_____

_____

_____

Least Energetic: _____

Sound:

Decibels = $10 \log\left(\frac{I}{I_0}\right)$

Decibels is how loud sound is.

$I_0 = 1 \times 10^{-12} W/m^2$ Threshold Intensity of Human Hearing. The lowest we can hear.

Intensity = $I_0 10^{\frac{dB}{10}}$

Loudness of Sound:	
20 dB	Very faint
40 dB	Faint
60 dB	
80 dB	Loud
100 dB	
120 dB	Very Noisy and Painful
140 dB	
160 dB	Unbearable

543

Final Exam Review

Name_____Period_____

1.... A 10.0 Kg box is pulled at an angle of 40° with a force of 110 N and $\mu = 0.05$. Find the following:

B) Weight:

B) Normal Force:

C) Force of Friction:

D) Net Force:

E) Acceleration:

F) What is the displacement after 1.00 minute?

G) What is the work done in 1.00 minute?

H) What is the Power delivered in that time?

2...A 2.80 kg box is pushed with a Force of 560 N. The Coefficient of Friction is 0.9. Find the following:

A....Draw a Free Body Diagram for the Box:

B...What is the Acceleration of the box?

C...What is the time taken to cover a distance of 200m?

D...What is the Work done?

E...What is the Work done by Friction?

3...Indicate the direction of the Restoring Force in each part of the oscillation below:

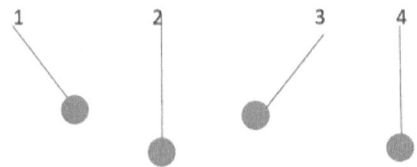

Indicate the locations where the Restoring Force is zero:

Indicate the locations where the restoring Force is Max Right:

Indicate the locations where the restoring Force is Max Left:

Indicate the locations where Velocity is Max:

Indicate the locations where Velocity is zero:

Indicate the locations where the Oscillation is at Equilibrium Position:

4... In the roller coaster below friction is considered negligible

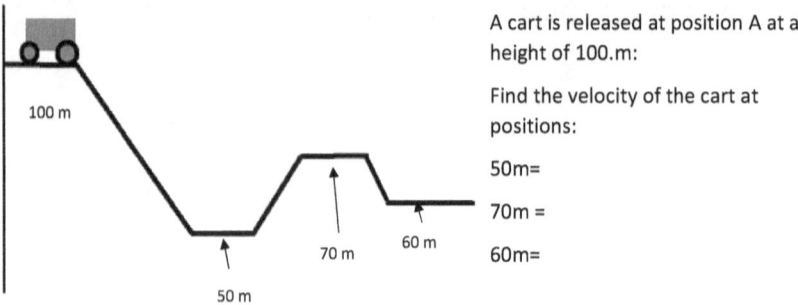

A cart is released at position A at a height of 100.m:

Find the velocity of the cart at positions:

50m=

70m =

60m=

5...If the cart in problem 4 has a mass of 20.0 kg what is the Total Energy of the System?

6...What is the First Law of Thermodynamics?

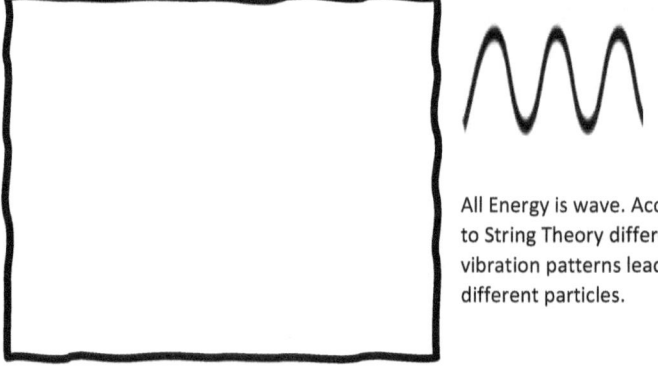

All Energy is wave. According to String Theory different vibration patterns lead to different particles.

Label the locations in the Oscillations:

7…State where is in the Oscillation of a Spring is the Kinetic and Potential Energies maximum or zero:

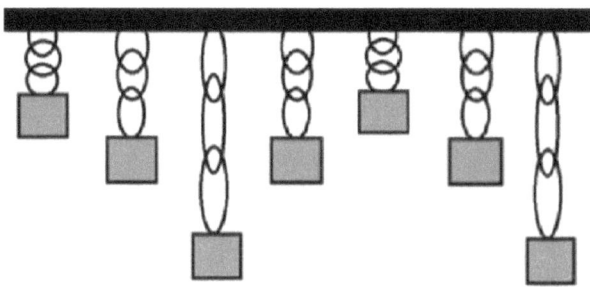

Max Potential Energy:

Max Kinetic Energy:

Zero Potential Energy:

Zero Kinetic Energy:

8...A person applies a variable force as shown in the graph below. Answer the following:

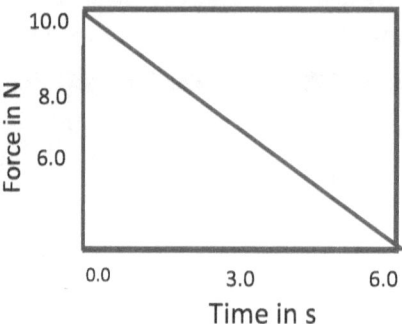

A...What is the Impulse on the box?

B...If the box is 3.2 kg what is its gain in Kinetic Energy if there is no Friction?

C...What is the Final Velocity?

D...What is the Average Acceleration?

E...What is the Power delivered to the box?

9...Below are Standing Waves in a Guitar String that is 1.00 m long. The Speed of the Wave is 80 m/s. Fill the table below:

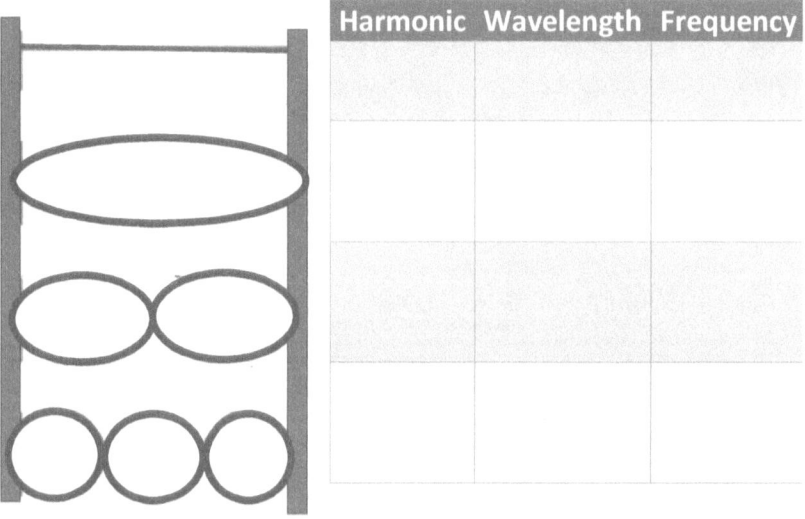

Harmonic	Wavelength	Frequency

Harmonic	Number of Nodes	Number of Anti Nodes

10....Fill out the Table below:

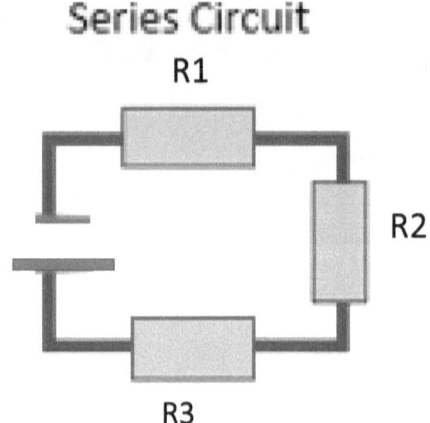

Resistor	Voltage	Current	Resistance	Power
1			10.0 ohms	
2			10.0 ohms	
3			8.0 ohms	
Total	60.0 V			

11...Third Table

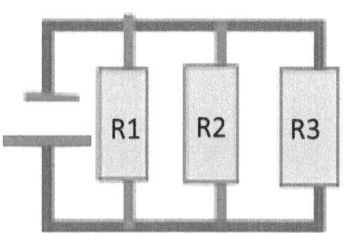

Parallel Circuit

Resistor	Voltage	Current	Resistance	Power
1			11.0 ohms	
2			9.0 ohms	
3			13.5 ohms	
Total	9.0 V			

12...Draw an arrow for the direction of the Magnetic Field produced in these Solenoids (Coils) with current in the direction shown:

13...Draw a Force Vector for the Wires carrying Current in the following directions

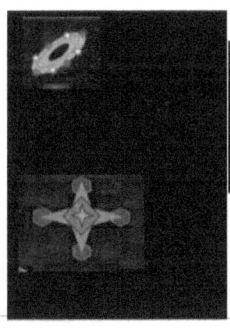

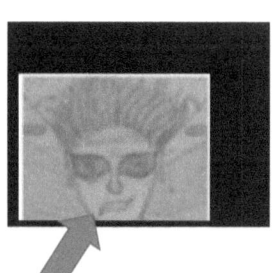

Space is filled with Electric and Magnetic Fields:

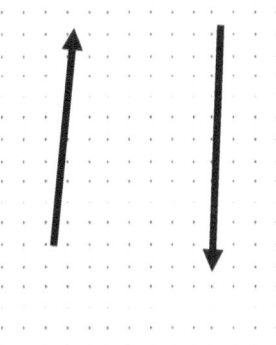

x x x x x x x x x x x x x x x x x x
x x x x x x x x x x x x x x x x x x
x x x x x x x x x x x x x x x x x x
x x x x x x x x x x x x x x x x x x
x x x x x x x x x x x x x x x x x x
x x x x x x x x x x x x x x x x x x
x x x x x x x x x x x x x x x x x x
x x x x x x x x x x x x x x x x x x
x x x x x x x x x x x x x x x x x x
x x x x x x x x x x x x x x x x x x
x x x x x x x x x x x x x x x x x x

Game

Group Name_____Period_____

Students_____

_____

Conquering the World

The goal of this game is to conquer parts the world by getting an answer correct. The first of the four teams to get the right answer for a question wins that specific territory. If there is a tie, like when two or more teams answered at the same time, an electronic coin is rolled to see who conquers it. The game ends after the last question. If no team is able to answer a question, that territory is left blank unconquerable. A team wins the game if it has conquered the greatest number of territories. A tie is broken by rolling a coin.

Map:

North America	Artic	Europe	Russia	China	Japan
Mexico	Africa	Egypt	Middle East	India	
Brazil				Australia	
Argentina	Antarctica				

1....North America:

What is the change in Kinetic Energy of an object moving at a Speed of 10.0 m/s?

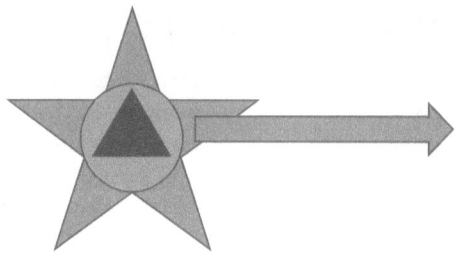

2...Mexico

What is the Work done on a 3.0 kg Particle if its Impulse is 150 kg m/s and it Accelerates from 10.0 m/s to 60 m/s?

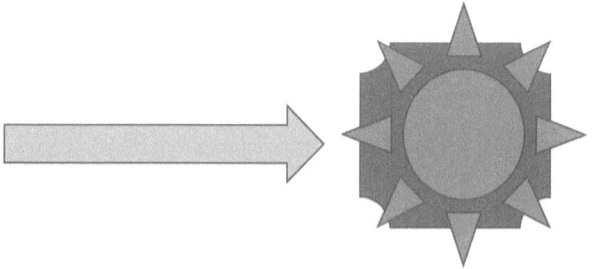

3...Brazil

How long does it take to push a 10.0 kg box a Distance of 64.0 m if its Final Velocity from rest is 10.0 m/s?

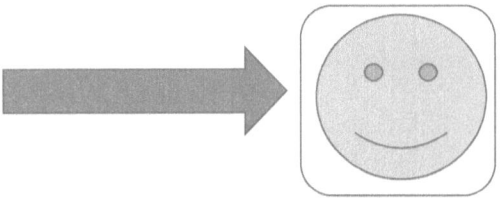

4...Argentina

What is the Power put on a 9.00 kg Particle in 7.0 s if it Accelerates from 6.00 m/s to 7.00 m/s?

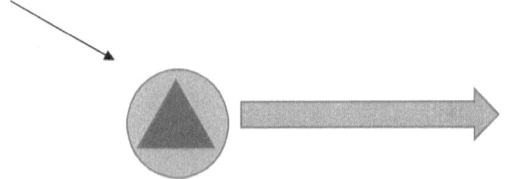

5...Antartica

What is Acceleration due to Gravity on the Surface of a Planet if its Gravity is 1/6 of ¾ of the Earth's Gravity?

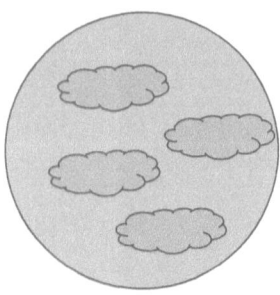

6...Artic

How many Earth Days does it take Planet Yur to revolve around its Sun if their year is 8 times longer than the Earth's Year?

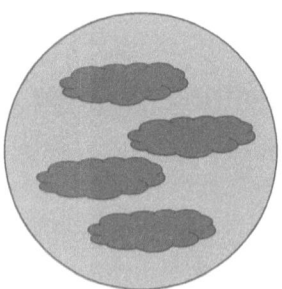

7...Europe

A Wave oscillates at 90 Hertz. What is its Angular Frequency?

8...Africa

How long does it take a 90 Hertz Wave to complete one oscillation?

9...Egypt

What is the Length of a Pendulum for its Period to be exactly 1.00 s?

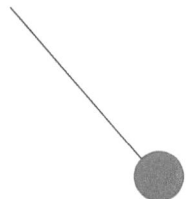

10...Middle East

What is the Period on a Spring if its Spring Constant is twice the Hanging Mass?

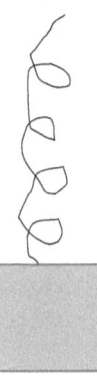

11....Russia

What happens to the Period of Oscillation of a Spring Mass System it its stretched twice more and allowed to Oscillate?

12...India

What is the Length of a Pendulum that has the same Period of a Spring of K = 9.0 N/m and with a 6.00 kg mass Hanging on it?

13...China

What is the Gravity on the surface of a Planet if a 1.0 m long Pendulum oscillates with a Period of 1.2 s?

14....Japan

What is the Period of an Oscillation with Angular Frequency of 1.00 Rad/s?

15...Australia

What is Frequency of an Oscillation it its Period is twice the Period of Earth's Revolution around its axis?

The Group that was able to conquer the greatest amount of territories in the world wins. If there is a tie, use a dice or a coin.

Challenge 1:

A box is Accelerated with a Spring of k = 9.0 N/m that is compressed 0.10m. The box has a mass of 1.0kg and after released by the Spring it slides over a frictionless surface and then falls off a cliff that is 80.0 m high.

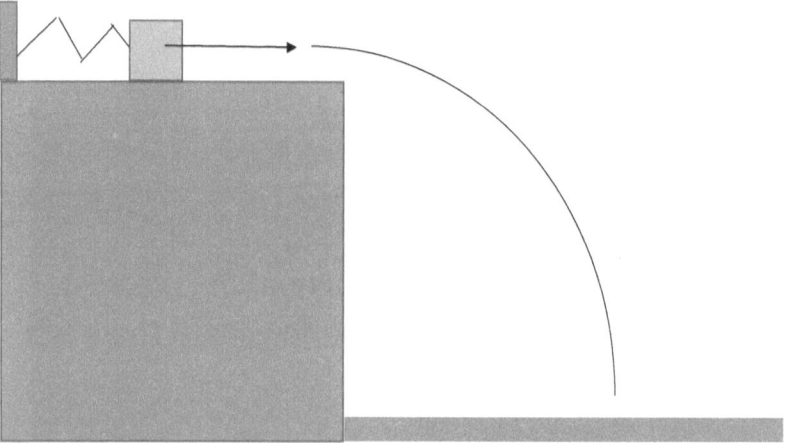

A...What is the Velocity of the ball after it is released by the Spring?

B...What is the Range of the box after if it falls off the cliff?

Challenge 2:

A ball first falls a Height of 200 m in the absence of Air Resistance. Each time the ball hits the ground it loses 7% of its Total Energy. How many times must the ball bounce in order to reach a height of only 104 m?

Challenge 3:

What is the Period of a 1.00 m long Pendulum inside an Elevator Accelerating down at 7.00 m/s^2?

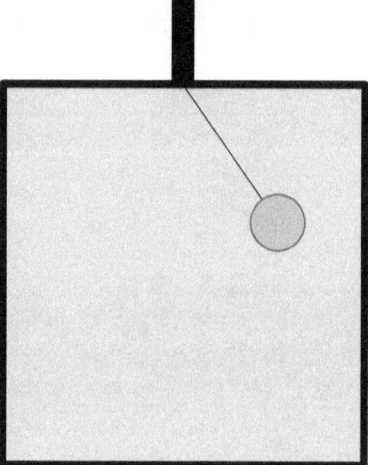

What is the Angular Frequency?

What is the Total Acceleration pulling the Pendulum down?

About the Author

I, Diogo Franklin de Souza, was born in the city of Rio de Janeiro, Brazil in August 20, 1986. I moved to Dallas, Texas when I was 11 years old. I write stories since I was 9 years old. My books tend to contain short summaries of the most important things I find about life, morality, philosophy, and science. Like I say, everything is part of a whole system, and this is also for everything I do and write. I always wanted to have all the most important knowledge in only a few short books. That is why I write, and that is my inspiration for short summaries. I hope this book brings some inspiration also for the readers, because that really is the purpose of my work. Read it and take from it, pieces of gold for you that can be useful in your life. Enjoy....